Land of Nuclear Enchantment

Lucie Genay

Land of Nuclear Enchantment

A NEW MEXICAN HISTORY OF

THE NUCLEAR WEAPONS INDUSTRY

University of New Mexico Press
Albuquerque

© 2019 by the University of New Mexico Press
All rights reserved. Published 2019
Printed in the United States of America

Library of Congress Cataloging-in-Publication Data
Names: Genay, Lucie, 1987– author.
Title: Land of nuclear enchantment: a New Mexican history of the nuclear weapons
    industry / Lucie Genay.
Description: Albuquerque: University of New Mexico Press, 2019. |
    Includes bibliographical references and index. |
Identifiers: LCCN 2018024375 (print) | LCCN 2018036459 (e-book) |
    ISBN 9780826360144 (e-book) | ISBN 9780826360137 (printed case: alk. paper)
Subjects: LCSH: Nuclear weapons industry—New Mexico—History. |
    Nuclear weapons industry—Social aspects—New Mexico. | Nuclear weapons
    industry—Environmental aspects—New Mexico.
Classification: LCC U264.4. N32 (e-book) | LCC U264.4. N32 G46 2019 (print) |
    DDC 338.4/76234511909789—dc23
LC record available at https://lccn.loc.gov/2018024375

Cover photograph courtesy of Shutterstock
Designed by Teresa W. Wingfield
Composed in ITC Galliard

Contents

Foreword

In the popular imagination, the world's first nuclear weapon was designed and assembled on a mountaintop laboratory shrouded in secrecy, tested in a remote desert too far away from human habitation to affect anyone, and located in an isolated, vaguely foreign region known as the state of New Mexico. From the beginning, the nuclear enterprise was portrayed as a heroic undertaking. In this saga, the protagonists were larger-than-life figures such as J. Robert Oppenheimer, a brilliant physicist and leader of the scientific team, whose testimonial at the first atomic blast was an allusion to the Bhagavad Gita: "Now I am become Death, the destroyer of worlds."

The New Mexicans most affected by these developments were not likely to have quoted Hindu scripture. In their daily practice, some would have turned to the nearby sacred peaks, Tschicoma or Redondo, and invoked Indian deities for rain and for healing. Others would have prayed following the traditions of folk Catholicism. The Los Alamos laboratory was located on the eastern slope of an immense volcanic rim, and to the outside observer the region may have seemed sparsely populated. But as the eagle flies, Los Alamos is only a moderate distance to several pueblos (Jémez, Cochití, San Ildefonso, Santa Clara, and Ohkay Owingeh), to a like number of Spanish or Mexican land grants (Cañón de San Diego, Ojo de San José, Baca Location No.1, Ramón Vigil, Juan José Lobato, Cañada de Santa Clara, and Abibuiú), and to the many small villages that radiate from the town of Española in all directions. These sacred mountains and communities would have been adversely affected by a nuclear accident at

Los Alamos, and some did face serious hazards because of the dumping of radioactive and chemical wastes along the watershed.

This book focuses on how New Mexicans, specifically Native peoples, *nuevomexicanos,* the ranchers of the Tularosa Basin, and the uranium miners of Grants experienced the establishment and expansion of the nuclear weapons industry in the state. These New Mexicans found their lives upended and transformed, even when they were not employed by the labs or directly involved in the nuclear enterprise.

Without articulating a formal theory of settler colonialism for New Mexico, Lucie Genay notes the parallels between these events and the region's colonial history: the subordination of Native peoples by the Spanish; the conquest of the territory by the US Army in 1846; the postponement of New Mexico statehood for more than sixty years; and the loss of land and folk traditions associated with the imposition of neocolonial status. The expropriation of private land by the federal government in the 1940s differed from these previous experiences in significant ways, but nevertheless represented a new form of conquest.

With the Manhattan Project, scientists, engineers, and federal dollars poured into New Mexico and created—directly and indirectly—thousands of blue-collar jobs in a region dedicated to agriculture and burdened by rural poverty. Economic opportunity, however, came at a substantial cost. Oral histories recount searing experiences of ethnic and gender discrimination. The enterprise was launched in haste and the subsequent lack of oversight resulted in the indiscriminate release of toxic wastes, threatening land and contaminating water critically important in an arid environment. The radioactive fallout from the Trinity test adversely affected the health of residents downwind. Homesteaders and ranchers abruptly displaced by the federal takeover of the Pajarito Plateau and White Sands lost a source of livelihood during the war that they believed was temporary. They subsequently discovered that the federal occupation was permanent. This is a complex story that requires sound judgment and careful attention to detail.

Genay tells this story with the sensitivity of someone attuned to the significance of cultural and class difference and to the weight of historical tradition and a people's attachment to land. Based on a large collection of oral histories and a wide range of documentary sources, this book narrates with verve the history of the "scientific conquest" of New Mexico. Indeed, that conquest, broadly based and far-reaching in its consequences, is the singular watershed event of New Mexico since statehood.

If to the rest of the world the Pajarito Plateau, White Sands, and the state of New Mexico seem less remote today, it is largely because of the Manhattan Project and the role of this region in the launching of the nuclear age. Reading this volume gives us an excellent opportunity to evaluate both history and contemporary meaning.

L. M. García y Griego
CAÑÓN DE CARNUÉ, NEW MEXICO

Acknowledgments

My thanks first go to the University of New Mexico Press for giving a young French researcher the incredible opportunity to publish her work and make it available to the people to whom it is addressed. I am grateful to many who helped and supported me during my research in New Mexico; their cooperation, their assistance, their counsel, and their friendship brought this project to a successful conclusion. I would like to particularly thank Dr. L. M. García y Griego, whom I met when I was an exchange student at UNM from 2008 to 2009. He has become my mentor and friend. His enthusiasm, insight, and encouragement were invaluable. I also wish to heartily thank Dr. Gerald Vizenor, whose course on the Manhattan Project inspired me to work on the subject. He was among the first to offer advice on my project. Other UNM professors were involved in my research. Although I cannot name them all, I express my gratitude to all of them.

For their professionalism, kindness, and precious knowledge of archival collections, I would like to thank the staff of the Center for Southwest Research. For the wonderful working conditions they offer visiting researchers, I would like to thank all UNM libraries. I deeply appreciate the helpfulness of all who assisted me at the various facilities and archive centers I visited throughout the state during my research trips. Most important, I am indebted to all the New Mexicans I met and to whom I presented my work; those who agreed to share their opinion and experience with me, as well as those who explained why they would rather not. Too many offered their help to name here, but they are all represented in one way or another in these pages. For their friendship, their precious ideas, and, for some of them, their company on my trips, I would like to thank my friends from the "Land of Entrapment." Special thanks go to my wonderful proofreaders on both sides of the pond.

On the French side, I am particularly grateful to Dr. Susanne Berthier-Foglar, my advisor since my third year at university, who has been supporting me unfailingly throughout my academic career. I am also grateful to my colleagues at the University of Limoges for their warm welcome and kind support through the final stages of this process. Finally, this book would not exist were it not for the love of my family and friends, who have been there for me every step of the way and helped me be positive in the most difficult times. My thoughts particularly go to my dear Mamema. *Un grand merci à toutes et à tous.*

Abbreviations

AEC	Atomic Energy Commission
AFB	Air Force Base
ALAS	Association of Los Alamos Scientists
ALO	Albuquerque Operations Office
AT&SF	Atchison, Topeka, and Santa Fe Railroad
AT&T	American Telephone and Telegraph
BTL	Bell Telephone Laboratories
CCC	Civilian Conservation Corps
CLER	Citizens for LANL Employee Rights
CCNS	Concerned Citizens for Nuclear Safety
CIC	Counter Intelligence Corps
DOD	Department of Defense
DOE	Department of Energy
EIS	Environmental Impact Statement
ENDAUM	Eastern Navajo Diné Against Uranium Mining
EPA	Environmental Protection Agency
ERDA	Energy Research and Development Administration
FAS	Federation of Atomic Scientists
FBI	Federal Bureau of Investigation
GM	General Motors
GNEP	Global Nuclear Energy Partnership
GTCC	Greater-Than-Class C
HOME	Healing Ourselves and Mother Earth

ICBM	Intercontinental Ballistic Missile
ICRP	International Commission on Radiological Protection
IFNEC	International Framework for Nuclear Energy Cooperation
ISL	In Situ Leach
LAHDRA	Los Alamos Historical Document Retrieval and Assessment
LANL	Los Alamos National Laboratory
LARS	Los Alamos Ranch School
LASL	Los Alamos Scientific Laboratory
MASE	Multicultural Alliance for a Safe Environment
MED	Manhattan Engineering District
MIT	Massachusetts Institute of Technology
NASA	National Aeronautics and Space Administration
NCRP	National Council on Radiation Protection and Measurements
NPL	National Priorities List
NMED	New Mexico Environment Department
NMSM	New Mexico School of Mines
NMSU	New Mexico State University
NNSA	National Nuclear Safety Administration
NRC	Nuclear Regulatory Commission
NTS	Nevada Test Site
OFCCP	Office of Federal Contract Compliance Programs
ORNL	Oak Ridge National Laboratories
R&D	Research and Development
RECA	Radiation Exposure Compensation Act
REM	Roentgen Equivalent Man
SM	Staff Member
SNL	Sandia National Laboratory
TEC	Technician
TERA	Terminal Effects Research and Analysis group
UNM	University of New Mexico
URI	Uranium Resources Inc.
WIPP	Waste Isolation Pilot Plan
WPA	Works Progress Administration
WSMR	White Sands Missile Range
WSPG	White Sands Proving Grounds

Land of Nuclear Enchantment

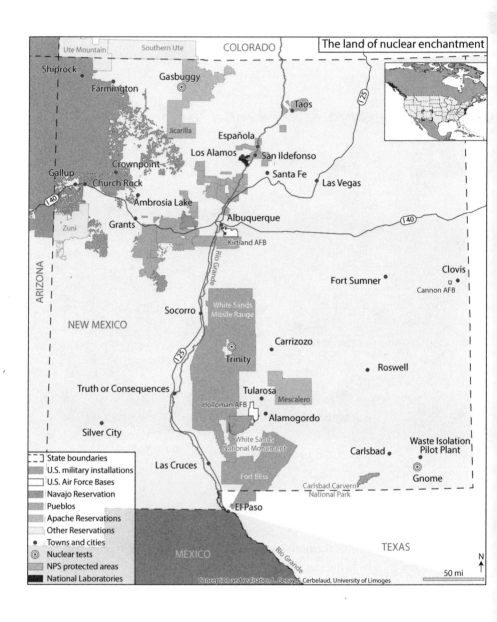

The land of nuclear enchantment

COLORADO

Ute Mountain   Southern Ute

Shiprock
Farmington
Gasbuggy
Jicarilla
Taos
123
Española
Los Alamos   San Ildefonso
Crownpoint
Gallup
Church Rock
Santa Fe
Las Vegas
Ambrosia Lake
Zuni
Grants   Albuquerque
140
Kirtland AFB
Rio Grande
Clovis
ARIZONA
Fort Sumner
Cannon AFB
NEW MEXICO
Socorro
White Sands
Missile Range
125
Carrizozo
Trinity
Roswell
Truth or Consequences
Tularosa
Mescalero
Holloman AFB
Alamogordo
Silver City
White Sands
National Monument
Waste Isolation
Pilot Plant
Carlsbad
Las Cruces
Gnome
Fort Bliss
Carlsbad Cavern
National Park
El Paso

MEXICO   TEXAS
Rio Grande

State boundaries
U.S. military installations
U.S. Air Force Bases
Navajo Reservation
Pueblos
Apache Reservations
Other Reservations
Towns and cities
Nuclear tests
NPS protected areas
National Laboratories

N

50 mi

Conception and realisation L. Genay, F. Cerbelaud, University of Limoges

One

Introduction: *Ground Zero*

At 5:30 a.m. on July 16, 1945, a futuristic weapon, dubbed the "Gadget," lit the sky in the Jornada del Muerto desert. On that day, New Mexico witnessed the birth of a new age and added to the yucca, the roadrunner, the piñon, the turquoise gem, and to the chili pepper, another state symbol: the mushroom cloud. The explosion climaxed a $2 billion military and scientific venture, the Manhattan Project, undertaken during the Second World War. Less than a month later, this new weapon practically annihilated the Japanese cities of Hiroshima and Nagasaki.

In the United States, the Manhattan Project is remembered as the enterprise that ended the war and saved the free world from fascism. The scientists who invented the bomb were celebrated as heroes. They had fought "the battle of the laboratories" in the shadow of secrecy, on the frontier of science.[1] The Project was history's most ambitious scientific undertaking—or "gamble," in President Truman's words, a mammoth enterprise that mobilized 130,000 workers across the nation.[2] Although the test on July 16 may not be remembered vividly today, its significance extends far beyond its wartime impact. The first atomic explosion ushered in an era in which human beings acquired an unprecedented ability to destroy, in which they dreamed of harnessing a seemingly inexhaustible energy source, and in which they transformed regions of the world into dominions of the nuclear industry.

In 1942, Manhattan Project sites were mushrooming in the four corners of the country. In the West, these sites increased the militarization of the region. They served as a vehicle for the federal government to weave a tight web of control over some long-discarded areas and formed the skeleton of what was to become a huge military, industrial, scientific, and academic complex. This complex was to be the future spearhead of America's nuclear strategy during the Cold War. I have chosen to term this conglomerate of installations the "nuclear weapons complex" to encompass all sites that are historically, financially, and politically connected to the manufacturing of nuclear weapons. New Mexico's fate within this complex was sealed on November 16, 1942. As it had been decided that the development, final processing, assembly, and testing of the bomb would be conducted in one place, J. Robert Oppenheimer, the scientific director of the so-called Project Y, suggested to his military counterpart, General Leslie Groves, that the world's first secret atomic laboratory be established on the location of the secluded Los Alamos Ranch School (LARS) on a mesa in the northern part of the state.

On February 7, 1943, the Manhattan Engineering District took official possession of the Ranch School in "the interests of the United States in the prosecution of the War," and waves of scientists and soldiers started to flood in.[3] Physicist John H. Manley called the arrival of the scientists "a new civilization colonizing" the Pajarito Plateau, "some 800 years after the first-known permanent inhabitants of this particular region, the Keres people."[4] At first, Project leaders estimated that three hundred people would be sufficient, but the population soon doubled, tripled, and quadrupled. By the summer of 1945, the community had reached six thousand people.[5]

At the war's end, Los Alamos experienced a mass exodus, as scientists were eager to put the barbed wire of the Army post behind them and return to their academic careers. The question the heads of the Manhattan Engineering District (MED) now had to answer was what would become of the laboratory on "the Hill"? New Mexico would be radically different had the Pajarito Plateau been restored to its initial isolation. Instead, Manley's "new civilization" continued its colonization. In the decades following the creation of site Y, the budding scientific community in the Jémez Mountains did not just grow to become the Los Alamos National Laboratories (LANL) and a ten-thousand-inhabitant town, it also spread, affecting the surrounding region.

The legacy of the Manhattan Project is both multifaceted and an ongoing process. Today, the tributaries of the current nuclear weapons industry can be mapped throughout New Mexico: two national laboratories, three test sites, three Air Force bases, waste disposal sites, nuclear reactors, nuclear weapon storage sites, the remains of uranium mining and milling, and White Sands Missile Range (WSMR), the country's largest military installation. The Laboratories were catalysts for the influx of scientific colonization, as they produced extensions and partner institutions throughout the state and impacted populations on both sides of the Rio Grande. This development flooded the region with new occupational opportunities and unprecedented demographic, socioeconomic, and environmental pressures.

The story I want to tell, therefore, is not based on a linear pattern like other histories of the Manhattan Project that begin with the genesis of atomic science and end with a successful scientific and military conclusion. It begins instead with a pattern of concentric circles emphasizing those closest to the epicenter. What impact did events of planetary magnitude, such as the Trinity test, have on the land selected for the experiment? How do the military project and its outcome fit into local history? New Mexico had experienced two major upheavals before: the Spanish conquest of the sixteenth century and the American conquest after the war with Mexico in 1846–1848. After World War II, the state underwent a phenomenal transformation with the arrival of what locals saw as an alien population of scientists and engineers—a third, scientific conquest. The previous colonization movements associated with the conquest of the Americas, it should be noted, were associated with scientific advances in transportation and warfare. In New Mexico during the Manhattan Project, however, the conquest was operated *in the name of* science and progress. This was in keeping with the American pioneering tradition of viewing progress as the greater good that justifies sacrifices, such as the environment, preexisting lifestyles, and cultural heritage. The scientific conquest led the young state on the path to becoming the nation's nuclear Eldorado, dragging along local populations who both benefited and suffered immediate and long-term consequences from this embrace with nuclear science. New Mexico, as the cradle of the nuclear age, acquired a new identity with far-reaching implications for its residents.

In the first decades of the nuclear era, the ready acceptance of a heavy, sometimes morally condemnable, price to pay for the advancement of civilization was abetted by tight secrecy. Theresa Connoughton, a Hispana resident of Santa Fe employed by LANL for twenty-one years before being laid off in November 1996, alluded to this condemnable price about which she had been "in total denial":

> You know, once this veil of denial is lifted, you have no choice but to fight it. . . . I tell you it is really an incredible picture to get when one finally realizes that one has been asleep for most of one's life. I think the science at the Laboratory is commendable, but, unfortunately, it has come at the sacrifice of people. And I think that, if people are sacrificed, then the product is useless.[6]

With the expression "sacrifice of the people," Connoughton refers to the stark inequalities between Los Alamos and the rest of the state as well as both ethnic and gender discrimination at the labs. The emotional aspect of her discourse conveys a sense of betrayal and guilt. My point is to look at the socioeconomic and cultural mechanisms that may have led to changes in perspectives such as Connoughton's and can account for such emotional responses. In other terms, this book seeks to explain process as much as it describes outcomes. A first approach to the Manhattan Project narrative in New Mexico is to say that the nuclear industry generated an immense economic boom that benefited everyone. A second approach qualifies this all-positive analysis by emphasizing the plight of those affected negatively by the nuclear weapons complex. Mrs. Connoughton's testimony is a mixture of both these perspectives: "commendable" science versus "the sacrifice of people." Both narratives provide valuable elements of truth that will not be ignored here, but what this book offers is an examination of the process that allowed for the transition from a positive, even miraculous, perception of atomic science and its economic effect, toward a less favorable assessment of that history and a more sharply honed criticism of the complex's environmental consequences, ethnic and class discrimination, and sociocultural impacts. This will be done by examining the context of oral testimonies and other evidence of the socioeconomic transformation of New Mexico.

I attempt here to reexamine a history that has usually been addressed from an external angle. The Manhattan Project has been a fascinating subject for many writers as evidenced by the abundance of publications since the so-called Smyth Report, the first official account of the scientific and technical development of the bomb, released on August 12, 1945.[7] The 1980s, in particular, were a prolific decade.[8] It produced Richard Rhode's celebrated book, *The Making of the Atomic Bomb*, and numerous publications that centered on the decision to drop the bomb on Japan.[9] The global consequences of the bomb monopolized the attention of historians and kept other aspects of its legacy in the dark, such as the effects on populations and the environment at the Project's sites. The first publications on the history of Los Alamos were, for the most part, narratives of scientific and logistical developments by participants who wrote their personal memoirs. General Groves, the military director, was the first to do so in 1962. In 1980, scientists remarked that Groves's book was the only published firsthand account of the Project, and Lawrence Badash, Joseph O. Hirschfelder, and Herbert Broida proceeded to publish a compilation of lectures that had been organized at the University of California in 1975.[10] Other scientific testimonies followed.[11] At the end of the decade, scientists' wives collected the memories of women who had followed their spouses to Los Alamos and some of them wrote their own memoirs. This represented a departure from military and scientific accounts to a social history of the community. Historian Cynthia Kelly and the Atomic Heritage Foundation also put together the accounts of scientists, military men, witnesses, politicians, and historians in several publications.[12] Those drew a lively picture of the atmosphere during the war, but they did not include testimonies of New Mexican participants.

After perusing the extant literature on one of the state's most important historical developments, this absence strikes one as conspicuous.[13] My approach centers on the experience of those individuals, whose participation was practically lost in history, and on their experiences of the postwar evolution, rather than on the historical developments and figures that were involved in the major events of that era.[14] Local residents have contributed to the success of the Labs, to the profits made by the uranium industry, to the construction and maintenance of storage sites, to America's military supremacy, and to the advancement of science. Therein lies the historical value of their stories.

Following a revisionist trend of social history, new studies began to appear in the early 1990s regarding how the bomb affected local histories and memories. In *Trinity's Children: Living Along America's Nuclear*

Highway (1992), Tad Bartimus and Scott McCartney introduce their book as a product of "hundreds of interviews conducted" along Interstate 25 between Las Cruces, New Mexico, and Wyoming. They call it "a journey," "part love story, part oral history" that "feeds off the emotion and excitement of its people."[15] This study builds on that tradition and emphasizes the perspective of common people—workers, farmers, and ranchers—who also played necessary roles in the nuclearization of New Mexico, but whose voices were stifled by national defense policies and by the weight of worldwide historical debates.

Beginning in 1991, the University of New Mexico oral history project "Impact Los Alamos: Traditional New Mexico in a High Tech World" conducted interviews in the Española Valley among residents who had worked or still worked at LANL. At the core of the project was the fundamental question of whether science had benefited the area or damaged ancestral cultures and eroded local traditions.[16] "Crucial as Los Alamos' global impact has been, its influence on local people and communities has often been neglected," wrote the managing editor of the *New Mexico Historical Review*, in a special issue, "Impact Los Alamos."[17] Carlos Vásquez, director of the project, partly accounted for this gap by identifying some of the obstacles the project encountered. He notes that "many locals saw the study as controversial and potentially compromising. Since LANL is the largest employer in the area and pays salaries often three to five times what other local employers pay, there is a premium in getting and keeping a job at the Lab."[18] I personally observed the same reluctance. Sociopolitical pressures still make it delicate to ask people about the nuclear industry in New Mexico.

Using these testimonies, this book traces the history of the scientific colonization, taking into account experiences of those who have often been cast to the margins of past historical writings: longtime residents of the region—specifically the Hispanos, members of Native American communities, and Anglos whose ancestors had arrived later in the nineteenth century. This work seeks to give voice to the Hispanos and Native Americans of the Jémez Plateau, blue-collar workers of Los Alamos, the miners and residents of the Grants Uranium Belt, and the ranchers and farmers adversely affected by the federal taking of land in White Sands Missile Range and whose lives were upended by the Trinity test and the US government's reluctance to address the "collateral damage" of the work at the Range. In the 1940s, many local Pueblo Indians, Hispano farmers, Mexican immigrants, and Anglo ranchers extensively relied on agro-pastoral activities for sustenance and lived on the fringe of industrialized America. The lives of some of these long-term residents were drastically altered by the development of the nuclear economy and ensuing social changes.

A brief historical overview is essential to understand the magnitude of the state's metamorphosis. When New Mexico acquired statehood in 1912, it had been a territory for sixty-two years. However, Americanization of the region had been sluggish. Most of the population still lived in a secluded world organized in self-sustaining villages, a few mining towns, and the larger cities of Albuquerque and Santa Fe. The completion of the Atchison Topeka and Santa Fe (AT&SF) railroad in 1879 was perhaps the last transformative event in New Mexico before the Manhattan Project. The railroad had accelerated Anglo colonization, connecting the oldest (non-Anglo) civilization in the country to the vibrant, cosmopolitan cultures of the East, along with their markets and sources of capital. In 1850, 1,500 Anglos lived in the territory. By 1900, their number had reached 100,000, about half of the population. Historian Pablo Mitchell calls this upheaval "the ultimate agent of American modernity and imperialism."[19] It was the ultimate agent—until the arrival of the Manhattan Project.

The railroad boom greatly transformed towns, gave jobs to thousands of people, and had a tremendous impact on the social fabric. Yet, for many New Mexicans, land remained the prime measure of wealth. It still was the chief—sometimes only—source of livelihood. In 1930, nearly 60 percent of the state's labor force was employed in agriculture.[20] When owning land was insufficient to escape poverty, people turned to unskilled and manual wage labor. Family members often had to travel far to supplement their meager farm incomes. A tiny minority had access to more qualified professions. The Anglo and Hispano elites held managerial positions, possessed large land holdings, and could engage in trade and politics. The extractive industries were the second most important economic sector.

At the brink of World War II, New Mexico was one of the poorest and most secluded places in the Union, where small communities strongly clung to their ancient ways to survive, and where economic strains were at their worst. When Project Y was established at Los Alamos, the exceptional circumstances of war deprived Hispano and Anglo homesteading families of the livelihood provided by the land they owned on the Pajarito Plateau, land they had to give up to the government on short notice. With the loss of land, however, came the gain of employment opportunities right next door; a blessing for a selected few in surrounding villages. Families no longer had to separate for months while some left to find work in other states. The Project, through the Zia Company, the principal subcontractor to the Lab, hired profusely from the Valley to do maintenance work at the secret laboratory.

As early as 1944, Manhattan Project officials targeted other places in New Mexico to carry on their atomic enterprise. More remote, extensive, and "uninhabited" portions of land were needed to test "the Gadget." The Alamogordo Bombing and Gunnery Range proved most useful for that purpose. The area and its military installations, now known as WSMR, is the largest overland military test range in the United States totaling 3,200 square miles and still works in close collaboration with the Labs. In March 1945, Project Alberta was initiated to assure that the bomb would be a practical airborne military weapon. New locations were again needed to pursue atomic experiments. Four months later, the Z division was created and moved to the military-owned Oxnard Field in Albuquerque, near Kirtland Air Force Base. Today, this locale is the site of the Sandia National Laboratories (SNL), which is the second largest employer in the state after the public school system.[21]

The propagation of science-related installations translated into a huge economic boom, higher employment numbers, increased income, and massive demographical growth. Albuquerque's population of about seventy thousand in 1940 exceeded two hundred thousand by 1960.[22] In turn, the multiplication of jobs and the influx of money and people fueled the creation and prosperity of other businesses and institutions of higher education that serviced the nuclear economy. Then, in 1950, a Navajo Indian, Paddy Martínez, found yellow coating on Todilto limestone he picked up on the lands of the Santa Fe Railroad at Haystack Mesa. It contained uranium, the mineral most precious to America's nuclear weapons complex.[23] As a result of this discovery, Grants became the uranium capital of the world and a magnet for prospectors and mining companies. Locals were employed to mine the radioactive ore and turn it into yellowcake through milling. With the addition of uranium extraction, New Mexico's nuclear industry became a cradle-to-grave industry, extending from raw material to the storage of old weapons. Then, in 1999, the Waste Isolation Pilot

Plant (WIPP) near Carlsbad received its first nuclear waste shipment after a twenty-year political battle that divided antinuclear activists, concerned citizens, and Carlsbad's enthusiastic promoters who were eager to welcome a project worth eight hundred jobs close to their town.

The conclusion to this accelerated historical summary is indisputable: World War II and the Cold War utterly revolutionized New Mexico's economic, demographic, and environmental landscape. There is also no denying that the nuclear weapons industry shaped a new social order and had repercussions on cultural patterns, but the principal omission that needs to be rectified concerns the repercussions of these changes on New Mexico residents. What did their home state's transformation signify for them and how did they respond to it? How can one evaluate the durability and distribution of the benefits entailed by the new booming industry—jobs, federal funding, attractiveness of the state, inflow of tourists, and so forth— and most important, to what extent did these populations gain from the high-technology revolution? What price did they pay? Aggregate statistics indicate a formidable rise in wealth and living standards, but in 2008 New Mexico still ranked fifth in the nation for the number of persons below the poverty line with its poverty concentrated in the state's Hispano and Native populations. This suggests they were not the principal beneficiaries of the scientific revolution.[24] Who was granted access to the high-paying jobs? Only by exploring these questions can one understand the wide range of opinions in New Mexico toward the nuclear industry.

Forty years after the Los Alamos Laboratory was declared to be a permanent installation, almost 70 percent of Los Alamos County's population was from out of state, a jarring number next to that of other northern New Mexican counties, such as San Miguel's 0.66 percent out-of-state population. Likewise, 3.5 percent of Los Alamos's population was under the poverty level, with a median family income of $30,307, compared to nearly 27 percent in San Miguel with a median family income three times lower ($10,841). At the same time, New Mexico ranked second in doctoral scientists and engineers per ten thousand in the nation, but twenty-ninth in number of high school graduates, and thirty-eighth in average annual income in the 1988 State Policy Data Book.[25] These figures suggest that the benefits of the nuclear industry were indeed distributed unequally among local and transplant populations and that, as a result, Los Alamos and the other nuclear complex–related areas had become privileged enclaves disconnected from the anxieties of surrounding communities.

To explain the stark inequalities in these statistics, this book looks at the history of the state's nuclearization and the ways that local communities were integrated into it. Early generations of Valley workers generally perceived employment at the Labs to be immensely advantageous. The salaries that locals earned enabled them to participate more actively in the growing consumer society. In this context of rapid industrialization and modernization, many agricultural workers were able to desert the rough and demanding labor on the farm for easier and more reliable work in the cities. Agricultural activities that had been the fragile pillars of the economy declined, accelerating the demographic shift from a rural, agrarian culture to an urban, industrial one.

Improvements in households were evident as people's purchasing power rose. Yet, many never caught up with the national standards of income. Somewhere on the course upward, something prevented them from going further. Priorities for New Mexicans shifted from survival to employment and education. Reliance on wage work built up a harsh competition among job seekers, and, in most cases, New Mexican workers and graduates from local higher-learning institutions were unable to compete with outsiders. Being granted access to the Labs or to other facilities of the high-technology corridor generated jealousies and tensions in the poorer parts of the state. Culturally, the nuclear economy opened a channel for dominant ways and ideas, some of which replaced the ancestral lifestyles based on the sacrosanct land. As in other instances of new industries entering traditional societies, cultural changes fractured the New Mexican identity, splitting it between pro- and antinuclear groups.

For those who succeeded in getting a job at the nuclear facilities, it meant entering a new culture, nuclearism. Psychiatrist Robert Lifton and Professor of International Law Richard Falk defined the term in 1982, stressing the irony behind the ideology, as a "psychological, political, and military dependence on nuclear weapons, [and] the embrace of the weapons as a solution to a wide variety of human dilemmas, most ironically that of 'security.'"[26] President of the Nuclear Age Peace Foundation David Krieger refers to nuclearism as "the belief that nuclear weapons and nuclear power are essential forms of progress that in the right hands will protect the peace and further the human condition"—the "right hands" being "one's own country" and "to further the human condition" meaning a "benefit to one-self, one's country, or one's corporation." Krieger identifies as key elements in the ideology that nuclear weapons are a necessary evil to keep the peace and that "the nuclear power industry is an absolute good."[27] These beliefs served the ambitions of the young Atomic Energy Commission (AEC) and supported the conviction that the country was under the threat of impending nuclear war. As a result, the testimonies of those who worked for nuclear facilities reflected these beliefs. Not all employees adapted the same way to the ideology, though. While some fully embraced it, others began to reject it.

Public awareness grew after the Three Mile Island accident of March 1979 and even more so after the collapse of the Soviet Union; both events dealt massive blows to nuclearism. In 1980, environmental reporter Phil Niklaus and writer Dede Feldman published a series of articles on the environmental impact of Los Alamos Scientific Laboratory[28] with the Southwest Research and Information Center, a leading actor in the rise of antinuclear activism in Santa Fe. "Solid waste materials, ranging in size from test tubes and rubber gloves to massive 'glove-boxes' and other laboratory equipment rendered useless by radioactive contamination, continue to be placed in huge trenches and shafts cut in the volcanic tuff at Los Alamos," they noted.[29] The dangerous dumping practices and unreasonable risk-taking in use since the war, which were the result of informed decisions based on the unsuitable safety standards of the time, had been maintained under the cover of secrecy. In the 1980s, revelations of the nuclear industry's most harmful environmental consequences greatly influenced local perceptions of the industry and further divided opinions. No longer protected by automatic government classification of their work, government-owned and privately contracted weapons labs came under attack by local activists who accused them of dealing irresponsibly with the dangerous by-products of their nuclear activities.

Supporters of nuclearism who argue that the rewards are worth the risk despite the evidence of the dangers of radioactivity continue to be a large group in New Mexico today. The state of New Mexico now has some of the most extensive weapons research, management, training, and testing infrastructures in the world. The downside to the sprawling industry is living in close proximity to nuclear weapons, toxic chemicals, radioactive materials and wastes, tests and experiments. With the half-life of radioactive materials being counted in thousands of years (twenty-four thousand years for plutonium), this legacy is an issue that will necessarily be passed on to future generations. Journalist Vincent B. Price published an alarming study of New Mexico's environment in 2011. In it, he sheds light on environmental catastrophes that, incredibly, went almost unnoticed, such as the Chuck Rock accident of July 16, 1979, when a dam on a huge evaporative tailings pond at a uranium mine burst, sending millions of gallons of radioactive liquid and tons of radioactive waste in the Rio Puerco.

Not only does Price's environmental study address one of the most concerning legacies of nuclear weapons production but it also emphasizes, once again, the unequal distribution of risk among populations. Price uses the term "nuclear colonization,"[30] which is another version of the term "nuclear colonialism" broadly used by indigenous peoples who denounce the use of their lands as nuclear dumping grounds. Longtime activist and Executive Director of Healing Ourselves and Mother Earth (HOME) Jennifer Viereck defined nuclear colonialism in 1992 as "the taking (or destruction) of other people's natural resources, lands, and wellbeing for one's own, in the furtherance of nuclear development."[31] Political authors and activists Ward Churchill and Winona LaDuke also refer to "radioactive colonialism."[32]

Why is it relevant to use a concept such as colonialism when addressing New Mexico's nuclearization? Historian of the American West Gerald Nash describes the West before World War II as "America's third world," an "underdeveloped area," "characterized by a colonial economy" where "agriculture, livestock, and mining were the major industries."[33] The description of prewar New Mexico above certainly fits this definition. Nash's use of the terms "colonial economy" and "America's 'third world'" helps underscore the idea that New Mexico had a subordinate relationship with the rest of the United States. It was dependent on the industrialized East, the prime market for its raw material production. The question is how New Mexico's postwar situation compares to this status and whether the state managed to free itself from outside economic forces thanks to the nuclear boon. Nash argues that the West broke free of its colonial bonds with the East thanks to the development of new industries; I will demonstrate that the case of New Mexico does not fit his conclusion.

In the process of changing its economic base, New Mexico's dependence on land and on eastern markets transferred to a dependence on government investments motivated by nuclearism. The exceptional circumstances of

World War II and the Cold War developed a federally sponsored economy. New Mexico's colonial relationship with the rest of the Union did not end but was extended and modified according to the nation's atomic objectives. People in the West had traditionally been wary of federal intervention, even though it had been substantial owing to the fact that western states were first territories under government and military control. One example was the building of public roads and highways. It was a win-win situation: "In aiding New Mexico residents, the government would likewise provide for its own needs. The opening of such roads would speed military operations in the area as well as decrease the high cost of shipping military supplies over inadequate routes."[34] The win-win concept was the same with the Manhattan Project, which was presented as serving the state's and the national government's interests simultaneously.

Historian Richard White defines the West as "a creation not so much of individual or local efforts, but of federal efforts" and as a region that "has been historically a dependency of the Federal Government." Citing the Bureau of Indian Affairs, the US Geological Survey, and the Forest Service, White argues that federal agents "were often more powerful than local political interests" and federal power could expand rapidly "because rival sources of political power in the states, local communities, and political parties were so weak." They were "conquered peoples."[35] His analysis establishes that colonialism was not only at the core of economic relations between New Mexico and eastern states, but it was also rooted in its political relations with the federal government. Sharing part of White's point of view, William DeBuys argues that the United States conquered New Mexico twice: first with "traders, miners, ranchers, and speculators—'rugged individualists' who served only themselves" and second with "soldiers, scientists, and other professionals who represented the United States as a collectivity."[36] The scientific conquest therefore fits into Debuys's second conquest (the collective representation of the United States) and into White's federal colonialism in politics.

As evidenced by these observations and activists' discourses, various types of colonialisms are blended in the scientific conquest. That is why I will use the term "neocolonialism" throughout this book. In the case of the uranium industry, for instance, corporate and government interests are both represented, so the extractive industry combines economic and government colonialisms. Economic colonialism differs from territorial colonialism in that the main actors are not nation-states but economic entities such as big corporations. The situation of economic dependence they create between themselves and the local population can be described in neocolonial terms. Neocolonialism thus is the most adequate concept as it refers to the economic and political policies by which a great power indirectly maintains or extends its influence over other areas or people.

Numerous elements in the state's relation with science can be spontaneously described using a neocolonial framework and lexical field. The concept of internal colonialism, specific to the United States, applies to New Mexico's association with science if one considers it as a story of encounter and exploitation between radically diverging populations who had had no previous contact and competed for the same resources. An external entity, in this case the federal government, imposed outcomes according to the logics of military priorities, acquired land, peopled it with settlers, exploited it economically, and extracted value from a local population for the benefit of its defense policies. Another element is the stigma of subordination of one people to another in a relation of political and economic domination that is often associated with ethnicity and that was also maintained despite economic growth.

These neocolonial relations have fueled profound, largely overlooked struggles among the inhabitants of New Mexico, between those who had been in the state for several generations before World War II and those who migrated to the state to take part in the nuclear boom. Even the creation of well-paying opportunities for workers from local villages is consistent with classical patterns of colonial enterprises, but the responses of the beneficiaries faced with the limits of this system will be discussed. Furthermore, the attitudes and mindset of both the colonized and the colonizers will be included in the analysis, as they were influenced by the state's sixty-two-year territorial history of external rule and wealth extraction. The varied consequences of the impact of the nuclear weapons complex are a unique opportunity to explore the different meanings and effects of neocolonialism.

For many New Mexicans, the longed-for proximity between home and work, the rise in incomes, and the hope of prosperity that came with nuclear science were seen as a windfall. But these shrouded pernicious, harmful repercussions. When the benefits of their entrance into a lucrative economy started to dwindle, locals gradually became aware of how much the new economy was costing them and would cost in the future, making them realize that perhaps they had struck a devil's bargain of sorts.[37] This book exposes the darker sides of an industry, much in the same way as does Hal Rothman's analysis of corporate tourism in *Devil's Bargains: Tourism in the Twentieth-Century American West*.[38] Rothman offers "a view of tourism from the perspective of the visited" and argues that the arrival of a tourist industry in areas or towns on the brink of extinction often results in a redistribution of wealth and power to outsiders, whom he contrasts with "neonatives." Very similar observations can be made regarding the nuclear industry, New Mexican "neonatives," and newcomers of the nuclear age.

As members of different ethnic groups have been affected in similar ways by the spread of the scientific conquest, it would seem illogical to focus solely on the ancestral Hispano and Native American peoples of the region. The prewar Anglo residents of the Tularosa Basin were also affected and they have been included in this analysis. By Anglo, I mean white US citizens who are not of Spanish or Mexican descent. By Hispano, I mean New Mexicans with Spanish surnames who share a common, principally religious and linguistic cultural heritage from their Spanish and Mexican ancestry. Being aware of the considerable differences among these groups, as well as among Native Americans, I attempt, when possible, to be more specific. I am, however, more interested in what these different groups of New Mexicans have in common or what could create tensions among them than I am in how to categorize them.

This narrative of the nuclear weapons industry is chronologically, but also thematically, divided into chapters that cover the main steps in the scientific conquest, its mechanisms, significance, and impacts. Chapters 1 and 2 provide a more thorough look at New Mexico prior to World War II, focusing on the preeminence of land and agricultural activities in its economy, culture, and history. Chapter 2 particularly addresses the steps leading to the arrival of science on the isolated Pajarito Plateau.

The following chapters revolve around the scientific conquest per se. After examining the land condemnation proceedings at Los Alamos and in the Tularosa Basin, chapter 3 focuses on the arrival of the first atomic scientists at site Y. A reexamination of these unusual pioneers' accounts shows how iconic images of the mythical West, the Frontier, and Manifest Destiny influenced their experiences and relations to locals. These images are contrasted with the memories of local workers, who saw the Laboratory first and foremost as unprecedented economic fortuity. Local witness accounts of the Trinity test follow, and the section ends with the pivotal decision to maintain, rather than dismantle, the laboratory at the end of the war.

Chapters 5, 6, and 7, the core of this narrative, detail the mechanisms behind the creation of New Mexico's nuclear weapons complex. First, the profits of big science in the state were exceptional and they generated hope that New Mexico might access its desired prosperity. The economic boom produced unwavering local support for the nuclear weapons complex. Second, the cooperation between big science and big government caused New Mexico to increasingly depend on federal funding to sustain economic growth. As a result, the local economy became tied to global events, particularly those impacting the national budget for nuclear weaponry during the Cold War. Third, secrecy was the mechanism that bound the other two together and affected New Mexicans the most: they were not aware of the dangers to which they were exposed.

The last chapters of this book deal with the expanding fallout—after the war and into the present—from scientific conquest. I address the environmental impact first, with an emphasis on notions of environmental justice and then dwell on socioeconomic and cultural implications. Last, chapter 10 is about the junction between New Mexico's atomic past and its complex future, a future that involves new parameters and includes new controversial questions on gender and race discrimination, on land compensation, on reorienting research away from nuclear weapons, on cleanup of contaminated sites, and on the opening of the Waste Isolation Pilot Plant. These topical issues are part of the inheritance of New Mexico's scientific conquest and prove that dependence on the federal government and on its nuclear weapons–related installations continues to be problematic in the state to this day.

Two

Land of Cultural and Economic Survival

"Official stories of the Manhattan Project tend to begin with scientific discovery and military necessity. The land and its people, their histories together, rarely appear, except perhaps as necessary sacrifices."[1] To reverse this, one ought to start with the land. Roxanne Dunbar-Ortiz affirms, "Everything in U.S. history is about the land."[2] New Mexico's history demonstrates the truth in that statement.

It All Starts with the Land

At the turn of the twentieth century, New Mexico's population formed a complex sociocultural collection of twenty-two Native American nations, descendants of Spanish conquerors and their mixed-race offspring, American frontiersmen and settlers, Mexican immigrants, and various other newcomers.[3] The status of New Mexico thus differs from other states where precolonial inhabitants were forced out or exterminated. Each period in the state's history melds together sequences of violent encounters with intricate cultural layering. The details of New Mexico's transitions from one sovereign to another are not relevant to this discussion. Rather, it is the way each transition contributed to the construction of a unique kaleidoscope of specific cultures that can provide the cornerstone for understanding the environment in which the scientific conquest materialized. Each new wave of conquerors that came looking for something they desired or needed (be it land, gold, or souls) altered the cultural dynamics of society. The scientific influx that began in the 1940s was no different.

Despite the successive waves of colonization and settlements, the diverse cultural groups of New Mexico consistently faced the same struggles as their ancestors: competing for limited resources of water, game, wood, and grazing lands. Environmental historian William DeBuys considers the competition "virtually as intense today as it was two thousand years ago, particularly if one adds 'employment' to the list of resources at stake."[4] Wage work as a livelihood came rather late in New Mexico compared to the industrialized East. Even after 1900, most kitchen staples were locally produced because villagers grew their food and bartered. Money was seldom used in transactions outside Albuquerque and Santa Fe where employment was easier to come by. The memories of interviewee Genaro Martínez from life in Chimayó before 1942 confirm that "there was no money," so each family "had to plant every corner that they had."[5] In fact, the absence of money was just as much the result of the reverse equation: there was no money *because* these families maintained self-sufficient lifestyles inherited from previous generations and based on mutual exchange and solidarity (people helped each other build their houses, Genaro said).

Likewise, Lebeo Martínez of Dixon explained that even though money was scarce and there were few signs of a consumer society (no cars, no pavement, no electricity, and no phones), there was "plenty of food because everybody had a big garden."[6] Armanda López Jackson (born in 1924), whose family had been in New Mexico for six generations, also spoke of a childhood setting of poverty but remarked that her mother would make do with a few milk cows and the food she grew on their farm (e.g., pumpkins, beans, potatoes) while the older siblings, who went away to work, would send money to buy items such as shoes for the younger ones. Armanda also recalled that there were no signs of "modern" development (cars or restaurants) in Rodarte when she was attending school with the nuns.[7] Poverty, endurance, and autonomy resulting from isolation are recurrent themes in their testimonies, but the lack of modernity is also an essential undertone, for it is a label that has stuck to the region and its people for centuries.

Despite cultural tensions throughout the territorial and statehood periods revolving mostly around land issues, which resurfaced during the Manhattan Project's tentacular expansion, New Mexico has consistently been "held up as a shining example of successful cultural *pluralism*."[8] Emphasizing cultural differences did not blur hierarchies, to the contrary. It had more "subtle" but "not less significant" colonial effects according to anthropologist Thomas H. Guthrie who argues that in New Mexico, the Native American and Hispano cultures "stand out, especially in comparison to Anglo-Americans, who often appear cultureless" and are therefore allowed "to occupy a normative position" as the "unmarked category," which "remains the standard or norm." Native Americans and Hispanos are thus "associated with 'tradition' and the past, leaving Anglo-Americans to claim 'modernity' and New Mexico's future for themselves."[9]

A striking illustration of this can be found in the West Wing of the Zimmerman Library on the campus of the University of New Mexico. The *Three Peoples Murals* completed in 1939 by Kenneth Adams depict each group's "contributions to civilization" as well as their union in the last panel:

> In the first panel the Native American cultural contribution is indicated by the arts, featuring basketry, jewelry, pottery, and weaving. The Hispanics in the second panel are engaged in agriculture and architecture. Anglo progress through science is the theme of the third panel. The final panel, The Union of the Three Peoples, looks to the future with Native, Anglo, and Hispanic united through a symbolic handshake.[10]

These paintings have been the subject of controversies and criticism because of their racist connotation. The last panel, especially, centers on an Anglo man who faces the viewer and looks ahead with blue eyes while the two other characters are painted from the side, their heads turned to the Anglo man. Their faces have no eyes as if they are blind and need guidance. Such paternalistic illustrations, which correspond to the thinking of the 1930s, conditioned and shaped the postwar development in many ways. The construction of the nuclear weapons complex markedly reinforced the perception of Anglo scientists as a "cultureless" group leading other New Mexican groups on the way to prosperity.

"Successful cultural pluralism" can be interpreted through the New Mexicans' shared attachment to their land.[11] For many New Mexican families prior to World War II, land meant *life*. Work was defined as *working off the land* in order to survive. Thus, the study of the preeminent significance of the land and the politics of land ownership in the region helps us understand how changes of the postwar period affected New Mexicans.

For Pueblo Indians, the people who came out of the earth, the land has a spiritual quality.[12] Pilgrimages to ancestral, sacred sites are a pillar of Indian culture, even after centuries of Catholicism imposed by Spanish conquerors. Paying homage to the land is a way of acknowledging connection to previous generations who do not, contrary to Christian beliefs, leave the earth to access another world (heaven or hell) but remain to reinforce the interconnection between people and their environment. The Pueblo people's sense of place and belonging is central to understanding how violent changes in the environment can influence their sociocultural structures. Anthropologist Edward H. Spicer depicts the Pueblo communities as "cultural islands" who managed to avoid extermination and acculturation.[13] It is indeed the cultural resilience and isolation of New Mexico Indians that has enabled them to retain their languages and their spiritual, artistic, and lifestyle traditions, and to endure historic upheavals over the centuries of invasion. At the beginning of the twentieth century, this population lived virtually the same way as their ancestors had.

Seminomadic nations—the Jicarilla Apache, Mescalero Apache, and the Navajo—also display their attachment to the earth in rituals such as the traditional burial of the newborns' placentas and umbilical cords in the earth as a symbol of a cyclic connection to the soil. Some nations are also known to have practiced geophagy, the eating of the earth. The permanence of the Navajo matriarchal system, which provides that property of land and livestock goes from the mother to the youngest daughter, despite being surrounded by a patriarchal Spanish-Mexican and then Anglo society, is proof of the capacity of the Navajo to use confinement and oral transmission as ways to preserve their cultural inheritance.

Hispanos are also considered to be the ancestral peoples of New Mexico. Their complex position between modern and traditional worlds is a key component in this narrative. While Hispanos do not consider land as central to their religion, there often is an attachment to it as something sacred, as the source of all life or the "mother and protector of their traditional subsistence pastoralism."[14] The Spanish colonists in the north "were true *paisanos*." One *Penitente Alabado*, a song from the Penitente Brotherhood, reflects a deep religious and cultural connection to the land: "De la tierra fuí formado, La tierra me da de comer; La tierra me ha sostenido, Y al fin yo tierra he de ser." (From the earth I was made, and the earth provides for me, the earth has sustained me, and at last earth I shall be also.)[15] Although represented differently than in Native spirituality, land is again at the center of life, as part of a cycle in accordance with Catholic funerary practices and New Mexican folk Catholicism.

Far from the centers of New Spain, the Mexican mainland, and the emerging United States, the Spanish enclaves of population "became bastions of cultural preservation, for they were at once so self-sufficient that they had little need for the outside world and yet so poor that the outside world had little need of them."[16] The Spanish settlements thus were similar to the "cultural islands" Spicer used to refer to Indian pueblos. Jacobo Romero from El Valle, a Hispano village in the Sangre de Cristo Mountains, emphasized the people's exceptional durability: "The people who wrung a living from Diamante's chilly meadows lived in many ways like the *indios* who gave the *banco del Apache* its name. . . . They depended for all they had or hoped for on the strength in their arms, their capacity to work, and *la voluntad de Dios.*"[17] This is something to which many New Mexico Hispanos still pay tribute. Journalist Juan Estevan Arellano considers Hispanos (he uses the term *nuevomexicanos*) to be in "communion with the landscape." Once "an alien presence" because of their Spanish forefathers, they "have now become as natural in this landscape as the piñón tree."[18]

The cultural ecology of northern New Mexico reveals that both Hispano villagers and Pueblo Indians relied on the land for virtually every aspect of daily life, be it food, houses, or art. Dry, floodwater, and irrigation farming techniques demanded labor and time, which was cadenced by seasons. Jacobo Romero recalled, "The net value of things was calculated not in dollars, possessions, or leisure, but in terms of food. . . . Meat was security, sustenance, the foundation of the household."[19] Those without land, on the other hand, would always be on the bottom rung of the social ladder and were the first to work as wage laborers (farm helpers, cattle herders, shepherds, and household servants).[20]

While farming had a religious significance in Native American cultures and while Hispanos considered crops and rain as the expression of God's generosity, Anglo settlers viewed agriculture differently. Anglo ranchers and farmers often aimed to participate in the market economy and expand their activities. Before the Mexican-American War, no more than a thousand Anglos were present in the region. After the war, New Mexico was traversed by all kinds of people who went through, came and left, but rarely stayed. One reason why New Mexico remained a territory from 1850 until 1912 even though it had reached the sixty thousand settlers threshold established by the Northwest Ordinance of 1787, was "the fact that the population was neither Protestant nor English-speaking."[21] The most aggressive critics of the territory did not hesitate to describe it as a burden on the United States. In July 1853, the military governor of the territory, General Edwin Vose Sumner (1851–1853), gladly left the region, concluding "that New Mexico was worthless for Anglo agriculture and industry" and "that New Mexicans were ignorant, indolent, and slothful."[22]

The Anglo conquest imported values centered on individualism, mercantilism, the cash economy, and private property, which often collided with the local self-sufficient Native American and Hispano cultures that generally relied on collective efforts to survive.[23] For the Anglo settlers, land was a commodity that could be exploited in capitalistic ways through ranching, mining, and cash-crop farming. Nonetheless, frontiersmen still owed their successes or failures to the land they had acquired through more-or-less honest tactics. Their entrepreneurial views partially explain why it took a while for a substantial, permanent Anglo community to establish residence in New Mexico. Simply put, workable resources were scarce and unprofitable. Ultimately, one characteristic of the land that transcended cultural schisms was the sense of autonomy that land ownership conveyed in a region cut off from the larger centers of "civilization." However, this stability was repeatedly challenged by changes in the legal definition of property that accompanied each conquest of the territory.

One key to understanding past and present struggles is the land grant system. These struggles are wholly part of the Manhattan Project's legacy because unresolved territorial issues reappeared years after the federal government took possession in Los Alamos of what used to be land grants from the Spanish crown to loyal citizens (there were 292 in 1848). Robert J. Tórrez, former New Mexico State Historian, comments on the manner in which the original grants were made: "You had to physically step on the land, run your fingers through the soil, and make a public commitment to live on it, cultivate it, and, if necessary, defend it with your life."[24] The ceremony, based on Spanish medieval traditions, suggests bonding with the land and the grants became important symbols of New Mexico's agro-pastoral past. The Spanish villages' social organization was centered on a collective exploitation of land and water combining both private and community grants. The arrangement enabled a relatively equal distribution of resources and preserved an indispensable natural balance for a sustainable exploitation of scant cultivable areas. Communal labor was used to maintain and regulate irrigation systems known as *acequias* (ditches).[25]

The Treaty of Guadalupe Hidalgo in 1848 states that "property of every kind now belonging to Mexicans . . . shall be inviolably respected."[26] Yet, the numerous violations brought about by the transition to the Anglo legal system stripped many original New Mexico citizens of their land and, consequently, of their livelihood.[27] Sociologist Clark S. Knowlton argues that New Mexicans "were left defenseless before the invading, dynamic, legalistic, lawless, and competitive Anglo-American civilization that did nothing to prepare them for adequate citizenship; stripped them of most of their land; reduced them to the situation of a conquered people without enforceable rights, and left them in extreme poverty."[28] The implementation of a colonial system between the territory and the Union was made possible through a system of land adjudication that favored the US government and American settlers over the local populations who, once deprived of the means of production, were left in a more easily exploitable position "as a surplus, cheap labor force, dependent on capital for their existence."[29] In addition, the land grab put an increasing strain on New Mexico's established cultural ecology, which was unfit for a market-oriented capitalist economy.[30] The practices of that period ushered a pattern for land appropriation repeatedly used thereafter by individuals, companies, and governments.

Industrialization is Late

The influx of Anglos had a far more visible impact on urban areas. Anglo merchants, lawyers, bankers, and politicians established their headquarters in Santa Fe. Even so, the town retained the atmosphere of a small community until the war. Rubén Montoya, who was born in Santa Fe in 1923 in a house on Jefferson Street built by his grandfather's ancestor about 350 years earlier, told the story of his grandfather who taught English and Spanish at the pueblos and of his grandmother who rented rooms near the famous Chili railroad line nicknamed after the *ristras* hanging on the front porches along the route. He described "a tightly-knit community" where people left doors unlocked and belongings outside and professed he "knew everyone in town."[31] That feeling waned after the war.

Further south, the emerging town of Albuquerque also grew with the flow of incoming Anglos. It became such a bastion of the Anglo influx that it changed its name from the difficult-to-pronounce "Alburquerque" with two *r*'s to the less challenging "Albuquerque."[32] While urban Hispanos owned small businesses and were discreet in politics, "hard-driving Anglos" made fortunes and invaded the political landscape.[33] Albuquerque split into two parts after the arrival of the Atchison, Topeka and Sante Fe (AT&SF) railroad in 1880. New Town undertook a more radical Americanization "through the replication of commercial and public architecture" found in the East and new big cities in the West.[34] The railroad became the largest employer in town boasting 2,500 workers and Albuquerque overtook Santa Fe as the territory's economic center.[35] Owing to terrain difficulties, the AT&SF line had bypassed the old capital.[36]

The railroad, a symbol of industrialization, changed aspects of New Mexican life elsewhere as it brought manufactured goods as yet unknown. Goods such as sewing machines, iron stoves, rice, raisins, and canned sardines appeared on the plazas of small communities like Chimayó, Córdova, or Truchas. In addition to gaining a right-of-way through the Native American reservations, the Santa Fe railroad employed thousands of Pueblo Indians, and job seekers followed the tracks to Arizona and all the way to California.

By 1910, the 1880 population of 120,000 in the territory had almost tripled. Nonetheless, more than half of the employed labor force was still engaged in farming activities. Agriculture remained the first sector until World War II despite a decrease in the 1930 census to approximately 41 percent of the employed labor force in the farm sector.[37] Ten years later, the majority of the New Mexican population still lived outside urban centers (355,417 out of 531,818). Albuquerque claimed thirty-five thousand people and Santa Fe twenty thousand.[38] While small communities mushroomed along the train tracks, others sprang up where coal, silver, gold, zinc, copper, potash, or oil was discovered. Not all mining communities survived the ups and downs of the extracting cycle. Some completely disappeared. Others managed to convert to other industries. The fluctuating nature of New Mexico's main markets calls attention to its precarious economic situation in the early statehood period.

Historian Hal Rothman explains how the region lagged behind the rest of the country in terms of industrialization as it relied on land to provide the industrialized East with raw materials that were then transformed and consumed there. Rothman writes, "No longer did mere ownership connote wealth and freedom. The value of land was predicated on the kind of products it could deliver to the markets of the East."[39] New Mexico's lateness in adopting the ways of the industrial age was in accord with its history of reclusiveness and its cultural distinctiveness. Roxanne Dunbar-Ortiz points again to the region's extended territorial period, which "allowed a continuance of *colonial-like* [emphasis added] conditions and uncontrolled economic exploitation of resources and labor."[40] That is why a dual, or even contradictory, picture can be painted of early twentieth-century New Mexico: massive changes were brought about by the growing Anglo presence but, at the same time, most of the state remained isolated from these changes and extremely poor by modern and contemporary standards.

The weight of agriculture, livestock, and mining in the local economy is the reason that Gerald Nash describes the West as an "underdeveloped" area, attributing New Mexico's poverty to the 40 percent of its population "of Hispanic origins" "clustered in small, rural communities," most of them "poor, engaged as they were in marginal farming" or "largely unskilled or semi-skilled workers" in towns.[41] Likewise, in 1955 Richard A. Bittman, in his thesis "Dependency and the Economy of New Mexico," observed, "Dependency is the very highest in the depressed agricultural regions where the activity is followed along cultural rather than commercial lines."[42] All living indexes computed by the Bureau of Agricultural Economics of the Department of Agriculture in the state since 1930 (e.g., farms with electricity, cars, and phones, and value of marketed products), for all but a few counties, were lower than the national and regional averages.[43] The correlation between ethnic background and poverty is correct. Correlation is not causal, but it encouraged attitudes that justified poverty as the result of ethnic origin.

Nevertheless, enduring seclusion and underdevelopment meant that poverty-stricken populations sought refuge in their cultural traditions. Ramón Frésquez, born in San Pedro in 1920, talked about his grandfather who homesteaded and his parents who raised cattle. As an altar boy, Ramón would go up to Valle Grande (great valley) with the priest, José Cubell of the Holy Cross parish in Santa Cruz, who went to say Mass for the boys Ranch School, which would later play a pivotal role in the history of the Manhattan Project. The family had a credit account at the [Boneno] grocery store in Española and would pay their bills with the chili peppers they produced and dried. He would also accompany his uncle to Taos to sell the apples and melons they grew. "There were no restrictions on the plaza or at the pueblos for selling things" at that time, he pointed out.[44] Nick Salazar, who was born in Chamita but grew up in San Juan (now called Ohkay Owingeh Pueblo), recounted how his family, who was in the sheep and cattle business, would also take the herd to pasture in the national forest (Valle Grande). Nick worked with them while studying at a Catholic school. He mentioned that one of the most notable traditional events for the community was the yearly fiesta.[45]

A solid network among families and communities was maintained through trade and barter. Florida Martínez (born in 1944) recalled these exchanges. When Florida was young, the people from Chimayó would come to Truchas to sell fruit that her mother would purchase and can. Meanwhile, her father made enough money to raise the children by cutting wood and other similar activities. About half of the Truchas population was able to stay in the village. Florida mentioned that she grew up speaking Spanish with her parents, but the authorities in mission school wanted students to speak English.[46] Language would be a major indicator of change throughout the twentieth century in New Mexican families. Joe Montoya (born in 1928) learned English at school but his father had become familiar with the language at his store thanks to the Anglos coming in. After working on the train in Denver, "not on the tracks, although many people did," Joe's father had found employment in Española as a clerk at the Bond Dweller store. Joe described Española as a beautiful place, which he never "had to" leave because his family's finances enabled him to stay. Interestingly, his uncle, Adolfo Montoya, lived in Los Alamos long before the Manhattan Project started there. He was a gardener for the Ranch School and Joe used to visit him as a boy when he was growing up in San Pedro, a place with few homes but many fields and orchards.[47]

Most interviewees considered automobile ownership as an important indicator of wealth. Nick Salazar recounted how he and his friends had to "pile" into a car or walk when they wanted to watch a movie in Española. Going to see a movie in itself was already a "big thing" for the youth in his community. Clothes were another marker of socioeconomic status: Nick owned a single shirt that he would wash every day. In Chamita, he said, there were neither rich nor poor families and only about four who had a few more material possessions than the rest (wagons, horses, and clothes).[48] Much in the same way, José Benito Montoya described Nambé, the community where he was born in 1929, as a poor village where children would walk to school and people used horses. His grandparents had come to the region about a 150 years before, his grandmother from Spain and his grandfather from Mexico. José Benito's father bought his first car for $25 after the war; it was an accomplishment. The vehicle could only take three passengers, and his father was not a good driver, but there was not much traffic in Nambé since he was one of only six people who owned cars. On the other hand, his parents never got a television set. In his memory, "a lot of people had to go out of state [for work] because there was nothing there."[49] Even though the influence of the dominant culture remained limited, increased communication led to more awareness that people had fewer possessions and led harsher lives in these communities compared to those living beyond the Valley. Class consciousness started to germinate.

By the 1920s, land exhaustion resulting from overpopulation, overgrazing, and overexploitation had become a serious problem. Money increasingly became indispensable for many New Mexicans, and the shift from the barter system to a cash economy confronted all villagers, not just the landless, with the need to find employment. Estimates before the Depression show that "some seven to ten thousand workers from villages in the Middle Rio Grande Valley left each year to work in sheep camps or to harvest crops in Colorado, Utah, Wyoming, Montana, and other western states."[50] Bernadette V. Córdova's family, for instance, led a migratory life. Bernadette is descended from the Vigils and the Montoyas, who had been in New Mexico for three generations. Every summer, her father, Gusmán Vigil from Sombrío, went to work alone on ranches and in potato fields in Colorado or in the mines of Arizona. His ten children went to pick tomatoes, prunes, and grapes in San José, California. Bernadette and her sister both started to work in the fields with Chinese and Filipino workers at age thirteen. They missed September and October at school. While they were at home, they would produce cereals (e.g., wheat, corn) and raise goats, cows, and chickens.[51]

Josefita Velarde's family also survived by living a migratory life. Josefita started school in Chimayó in 1917 before they moved to Colorado a year later to clear debts. The family moved to Albuquerque in 1919 and back to Chimayó the following year. Steady work was impossible, so they returned to the coal mines in Colorado until 1944 when she and her husband returned to New Mexico. Life was easier in Colorado, she stated, because they had electricity, stores, and good schools, whereas in northern New Mexico, people ate what they grew, they kept cows in the house for heat, and electricity came after the war. After this difficult childhood on the move, she was proud of her first job as a doctor's assistant at age nineteen. In her case, settling down was no longer the result of owning a parcel of land but of finding stable employment, a shift that many other families would eventually experience.[52]

One notes that the memories of local people who lived in northern New Mexico in the 1920s, 1930s, and 1940s are sometimes tinged with nostalgia for a time when life was simpler and closer to nature. But most expressed relief that these days are over, and children learned from the hardships their parents and grandparents had experienced to be grateful for the life-changing circumstances brought by the Manhattan Project. Nelson Kevin Vigil was born in Santa Fe after the war, in 1957. Like many others of his generation, his father had started working as a sheep herder in Wyoming at age eleven before working on the railroad at fifteen. He then became a barber and eventually found a job in Los Alamos. Nelson commented on his family's traditions, expressing his amazement at the old-timers' ability to make the most of the tools they had, such as a ten-mile ditch for irrigation. He stated, ironically,

> Now mobile homes are the only things that grow on fertile land, but at the time there was nothing to waste. They would terrace their land to grow food and they would build their house on the hill sides.
> . . . You had to show utmost respect to everyone, especially the older people. We would use *usted*. You had to be hard-working. There were rites of passage into maturity. For men it was when you would go work on the ditches alone for the first time with your shovel. For young women it was the first time they made a good tortilla.

Nelson's discourse tends to idealize the lives that his parents led because their sense of place and belonging was stronger. Their identity was based on their relation to the land as farmers, and their status within the community depended on their experience rather than on material possessions or what jobs they had. After the lab began employing members of these communities, the signifiers of status and authority gradually transferred to one's ability to get a good-paying job at the lab. The interviewee also talked of his experience in the eastern states when he joined the service and how the mindset he had formed through his Hispano education clashed with Anglo views:

> It's very infuriating to go back East and be looked down upon. They thought we were still part of Mexico, couldn't pronounce my name right. Called us Mexicans . . . Not that we are ashamed of our ancestors but we are Americans. The Hispanic mindset is "live and let live." It's not "go out and conquer," "grab anything you can." We take care of the land. I think the mindset has changed; people don't help each other as much.[53]

According to Nelson, previous generations seemed to have a clearer sense of identity because they defined themselves through their way of life, which they contrasted with the ways of conquering Anglos. At the same time as he puts forward his heritage and regrets its disappearance, Vigil expresses the desire to be considered an American, the equal of his fellow servicemen. This reaction reflects the struggles of many Hispanos and Native Americans within New Mexico's multicultural society, who have come to feel the downsides of belonging to "marked" cultures when the "unmarked" Anglo culture is considered the norm.[54]

Another factor, which relied on cultural exoticism, contributed to the decline of self-sustaining activities in favor of cash-producing pursuits in New Mexico in the first half of the twentieth century: the success of a burgeoning tourist market. Considering the region's economic difficulties and its historic reputation as a dangerous borderland on the edge of civilization, one may wonder who the first tourists were. Rather ironically, a substantial number of travelers came to New Mexico as a last resort. They were called "health seekers," "lungers," or "chasers" because they came to "chase the cure" thanks to the effect of the elevated, dry climate on respiratory diseases such as tuberculosis, then the leading cause of death in the United States.

Both in Santa Fe and Albuquerque, sanatoria and hospitals offering climato-
logical cures and hot spring treatments proliferated. Two of the most significant health migrants were J. Robert Oppenheimer and Dorothy McKibbin.
The former discovered New Mexico for the first time in 1921 when he came
to recover from dysentery and the latter sought treatment at Sunmount
Sanitarium in Santa Fe after she was diagnosed with tuberculosis in the winter of 1925. She said, "I fell in love with the place because of its beauty and
the cultural and intellectual atmosphere."[55] This atmosphere was due to the
many artists and writers who migrated to New Mexico for treatment. Nancy
Steeper, McKibbin's biographer, called it "an arts colony."

The Sanatoria movement and artistic magnetism launched New Mexico's potential as a tourist destination. Artists saw the Hispano and Native
American communal way of life as a poetic ideal that contrasted glaringly
with the suffocating industrial world of the eastern cities, ignoring the dire
economic difficulties that New Mexican farmers faced.[56] "Picturesque"
and "pristine" became the words most often used to describe the region.
English writer D. H. Lawrence, who visited the ranch that Mabel Dodge
Luhan had offered him northwest of Taos between 1922 and 1925, called
the state "the picturesque reservation and playground of the Eastern States,
very romantic, old Spanish, Red Indian, desert-mesas, pueblos, cow-boys,
penitentes, all that film stuff."[57] The same vocabulary would later be chosen
again by the first inhabitants of the Los Alamos scientific community to
describe their surroundings.

Meanwhile, the development of tourism was accompanied by the creation of national parks, proof of the growing presence of the federal government in the West.[58] This presence was the second "seed" that paved
the way for the postwar economic revolution (second to the development
of wage work and a cash economy). Several New Mexican interviewees
referred to the New Deal Civilian Conservation Corps (CCC) camps as
one of the limited job options they or other family members had in the
1930s. Along with national forests, parks, monuments—seven of them by
1933—military bases, and homesteads, these camps were the physical man-
ifestation of the increasingly conspicuous federal involvement in the state,
an involvement New Mexico became more accustomed to relying.

Some of the first "agents of the Federal Government" were not directly
employed by the government but were farmers. They were homesteaders
who increased "the value of raw land through construction of homes and

. . . cultivation of the land."[59] A few Hispanos also took advantage of the Homestead Acts to recover lost lands. In their case, however, homesteading was an activity of subsistence as opposed to "recreational homesteaders," idealists who wanted to get away from eastern cities and live a peaceful rural life and business farmers who saw homesteading as a path to economic gain.[60] Next to them, local New Mexican farmers appeared indolent and unable to profit from their resources. The reputation of New Mexicans as people devoid of the spirit of entrepreneurship enticed US officials to take a paternalistic position toward them and entrust maximum land to the safe care of the government.

That the resources of the West should be used for the greater good of the collective is a crucial precept to understand the postwar transformation of New Mexico. In the wake of the *U.S. v. Sandoval* case of 1897, an enormous swath of land—1.4 million acres—had been lost to the community, and the federal government had subsequently created the Carson and Santa Fe National Forests in the north.[61] To preserve what it saw as an eroding landscape, the Forest Service put restrictions on grazing seasons and herd numbers. While protecting the land, restrictions further jeopardized traditional lifestyles. Several interviewees indirectly alluded to these events when they talked about their families formerly taking cattle to the public domain of Valle Grande, the largest grass valley in the volcanic caldera in the Jémez Mountains. Ramón Frésquez mentioned his parents using the valley during the summers between 1925 and 1930, but by the time of the interview, in 1995, it had become private so "people needed a special permission or permit to go there."[62] Nelson Kevin Vigil also expressed regrets regarding permits people have to get from "the Game and Fish agency to hunt and fish. This is all subsidies for the government; it doesn't go back to the people of New Mexico. It doesn't make sense to me."[63]

When westerners, who traditionally distrusted the federal government, simultaneously experienced the Great Depression and a terrible drought in the 1930s, government action imposed itself as their best solution. The combination of these two factors immensely damaged New Mexico's fragile economy, intensified its vulnerability, and plunged the population further into poverty. Dependence on outside markets was emphasized when "in 1931, the state's most important crops were worth only about half of their 1929 value."[64] The value of minerals also dropped dramatically, affecting the lives of thousands of miners.[65] Plains dried out and ranches could no longer

feed cattle. In 1934, the Taylor Grazing Act aimed to control the degrada-
tion of public land and thus stabilize the livestock industry dependent on
those lands. All the ranches in the Tularosa Basin area fell under the category of degraded lands. According to former rancher David McDonald, the
drought "was of such magnitude that the government paid ranchers a dollar
a head for the privilege of putting their cattle to death."[66] Interestingly, the
Civilian Conservation Corps worked on Tom McDonald's ranch where the
Trinity "Gadget" was to be assembled a few years later.

Hipólita Fernández of Truchas points out that many people lived from
ranching before the war because the land seemed to produce much more,
and fields, livestock, and pasturage were much healthier. Her husband,
Delfido Fernández, was employed on a ranch, but like many others, he had
to work multiple jobs and was among those who found employment in the
CCC camps that had mushroomed throughout the state between 1933
and 1942.[67] Joe Montoya believed the CCC camps had been a positive
experience for his brothers who had left either for the camps or for the
war-related industries in California. He saw it all the more positively that
he minimized the impact of the Depression in the Española Valley. "It had
already been depressed before that," he said. The strained economic situation made no difference to them: "they didn't know any better."[68] Still,
New Mexico was among the last states in per capita income ($209 in 1932,
52 percent of the national average), and ranked first in both public illiteracy
(one-third of all schoolchildren did not attend class in 1933), and child and
infant mortality.[69] Unemployed New Mexicans were forced to survive on
government relief programs (60 to 70 percent of northern New Mexicans
by the mid-1930s).[70] That may have been the first time when a significant
portion of the New Mexican population relied on governmental help for
their livelihood. It was a harbinger of what was to come.

The New Deal policies would later work as precedent for the building
of federal institutions on New Mexican soil. Hal Rothman calls the New
Deal a "godsend" for the West because it offered "a backbone for tenu-
ous local economies and many opportunities for employment."[71] In many
ways, one could make the same argument for the arrival of the Los Alamos Laboratory. It was a godsend for the first generation of workers who
had just lived through the economic crisis and difficult climatic conditions.
They had come to see the federal government as a savior of economies
suffering from circumstances beyond their control.

The "Outdoor School" and Its Neighbors

With the onset of World War II, New Mexico finally had something the government urgently needed: isolation. One of the most isolated locations in the state was a place called the Pajarito Plateau. Nothing in the description of the 25-mile-long plateau at an elevation of 7,300 feet, 35 miles northwest of Santa Fe suggested suitability for the establishment of a high-tech company town that would eventually become the richest county in the state. Its remoteness was precisely what made it so attractive to Manhattan Project officials.

The history of the Plateau itself occurred in successive waves from the first Native American inhabitants to the Los Alamos Ranch School (LARS) and mirrors the state's stratified colonial history. The pueblo cliff dwellers, ancestors of the Keres people, were its first inhabitants in the twelfth century.[72] Under Spanish rule, Governor Gaspar Domingo de Mendoza (1737–1743) granted the land to a former soldier, Pedro Sánchez, from Santa Cruz de la Cañada in the Española Valley. José Ramón Vigil bought the grant in 1851, and the US Congress confirmed it in 1860. At the same time, Baca Location #1 was approved as a land grant in the Jémez valle area to the west where many local herders used to go with their sheep and cattle.[73] According to Stephen E. McElroy and Daniel Sawyer, who surveyed the Vigil Grant for the General Land Office of the Department of the Interior in 1877, the land was "valuable for its excellent grazing capacity and its large timber supply."[74] Subsequently, the grant was purchased by a lumberman from Oregon, Henry S. Buckman, who started exploitation of timber on the southern part of the plateau in 1898. His departure was followed by a five-year-long lawsuit (*Sánchez v. Fletcher*). Meanwhile, the US Forest Service acquired all the unclaimed land on the plateau, except the Vigil Grant, which was "set apart as a Public Reservation, for the use and benefit of the people." On October 12, 1905, Theodore Roosevelt, by presidential proclamation, named the area the Jémez Forest Reserve.[75] Ten years later, the US Forest Service established the Santa Fe National Forest by merging the Jémez and the Pecos National Forests. In the way it was handed from one owner to the next, both private and public, the plateau epitomizes the complexity of land quarrels in the Southwest.

Homesteading began on the Pajarito Plateau in 1887 when Juan Luis García of Guachupangue applied for a homestead and received his patent in

1892. In the end, there were about thirty-six homesteaders on the plateau, a majority of them Hispano farmers and ranchers.[76] The 1912 Homestead Act reduced the number of years required before requesting a patent from five to three years, thus increasing the number of applications between 1913 and 1917. While elsewhere in the country the homesteading movement declined, with the industrialist replacing the yeoman farmer as the epitome of America, on the plateau the homestead era lasted until World War II.[77] The 1912 Act allowed homesteaders to be absent from their homesteads for a period up to five months. The provision was especially convenient for Hispanos who instituted a system of seasonal migration on the narrow wagon road between the valley and the plateau twice a year. Sixteen out of the thirty-six homesteaders were the same people (or a relative) who would eventually have to sell their properties to the government after twenty-one to fifty-five years. The houses, cabins, cisterns, and other physical traces of these homesteaders' lives on the plateau were either integrated into the Technical Area of the new laboratory or were wiped out to make way for residential and recreational areas, such as a golf course, parks, roads, a library, or a cemetery.

In 1986, eighty-three-year-old Don Marcos Gómez shared his memories of homesteading on the plateau in a newspaper article titled "Los Alamos: He Lived on the Hill Before It Meant 'The Bomb.'" Nostalgically, Don Marcos said they led happier lives back then, though he acknowledged that his existence as a rancher, laborer, sheepherder, and cowboy had been an arduous one. He remembered that "the land was very fertile . . . and there was never a shortage of rain or *ojitos,* natural springs for watering livestock." They herded sheep and cows in Valle Grande, planted corn, beans, peas, potatoes, peaches, and apricots on the farm and then would go down the thirteen difficult miles to the valley with mules to sell some of their produce. "We'd buy the only things we couldn't harvest ourselves . . .—kerosene, matches, coffee, sugar." Wolves, bears, and mountain lions roamed the area. Gómez even talked of keeping rattlesnakes as pets. But what the old man emphasized was the solidarity and conviviality among families and the respect for elders.[78]

A few homesteaders were Anglo. Harold H. Brook had come seeking to recover from tuberculosis, but his ambition was to succeed as an agriculturalist. Therefore, "he brought with him an innovative, entrepreneurial spirit . . . and technical expertise far beyond that of his neighbors." He believed "that the region held the potential for more than subsistence."[79] He soon started acquiring other lands around him such as the Hopper and Benigno Quintana Homesteads and his mother, Mattie, applied for 150 acres upon his advice. Brook married Katherine Cross Brown of Santa Fe who is credited with choosing the name "Los Alamos" (the cotton woods) for her husband's ranch. Her son from her previous marriage, Frank Brown, would later be employed by the Zia Company, which, in 1946, was responsible for maintaining and operating the community of Los Alamos and the Laboratory facilities. In spite of his efforts, Brook could not make a viable business of his farm and became deeply indebted. He subsequently formed a partnership with Ashley Pond to manage the Pajarito Club. When their partnership failed, Brook sold his ranch to Pond and his mother's land to Edward Fuller, for whom Fuller Lodge is named. In 1917, Brook left the plateau and died of tuberculosis in Las Cruces, New Mexico.

Anglos generally had more difficulties than Hispanos who enjoyed the support of their community and could always stay with a brother, a sister, or a cousin in the valley in difficult times. To them, the plateau was not "a desolate and isolated place," but a "loose extension of their valley village life."[80] The Anglo ranchers, however, led a harsher, lonelier existence in the mountains. In contrast to the seasonal Hispano homesteaders, they occupied their properties year-round because they used the latest farming methods and had the technological means to irrigate, farm, and ranch throughout the year. This is probably the reason why Anglo owners succeeded each other rapidly. For all that, some of their names would become reminders in the atomic age of life on the plateau before the laboratory, while the names of Hispano homesteaders and other plateau families fell into oblivion for more than fifty years.

Apart from a few homesteads, the two main pre-LANL Anglo properties on the plateau were the Anchor Ranch and the Ranch School (LARS). Anchor Ranch originally was the 322 acres patented by Severo Gonzales and James Loomis. The land went from one owner to the next until Alexander C. Ross, who came from a wealthy New York family, bought it. Ross was mentally impaired and declared incompetent in 1915, so, although he was a permanent resident of the ranch, he did not personally oversee its operations, which were in the care of the Smithwick family.

In 1914 a group of industrialists from Detroit bought the Ramón Vigil Grant and founded the Pajarito Club. Harold Brook helped manage the estate until 1916 when he and Ashley Pond formed their partnership to build Pond's dreamed boarding school for boys after acquiring the other businessmen's share of the grant. The school opened its doors in the spring of 1917 and started to flourish under the directorship of Albert J. Connell, a former Forest Service ranger and scoutmaster in the Jémez Forest Reserve. The so-called outdoor school was designated under the category of "fresh air schools."[81] Most students were eastern urban boys with fragile constitutions who were sent by their wealthy families to discover the joys of nature and to toughen up thanks to the rough conditions of high altitude and outdoor activities built into the curriculum. Both Pond and Connell had been compelled to go to New Mexico because of their poor health and their personal experience with the unpolluted air of the plateau led them to believe it would have a fortifying effect on the boys' immune systems. Here is the description one could find in an advertising booklet Connell published in 1937:

> Los Alamos Ranch School is a six-form college preparatory school for forty-four boys 12 to 18 years of age. . . . Families from all parts of the country have sent their sons to the school. Graduates have prepared for and successfully carried on their college courses at Harvard, Yale, Princeton, Cornell, Williams, Amherst, Dartmouth, etc. A well-planned curriculum, an unusual diversity of valuable recreational interests, in a well equipped plant amid healthful and beautiful surroundings make Los Alamos particularly fitted to give a comprehensive education.[82]

It is interesting to note that the last page of the booklet features two pictures of Native Americans: one showing young Apache children ready for a fiesta and the other of Pueblo Indians ready for the Children's Dance at Cochití. The school thus drew a parallel between the education it provided and that of local natives, possibly capitalizing on the novel attractiveness of colorful, close-to-nature native peoples also advertised in tourist brochures. The school's concept relied on a form of nostalgia for a simpler, rural way of life inspired by New Mexico's environment and populations, but the time students spent there remained a parenthesis. They were not destined to lead such lives after graduation; a majority attended Ivy League universities and became prominent industrialists, businessmen, bankers, and lawyers.[83]

Providing healthy foodstuffs for students was an additional selling point. Due to its isolated location, the school relied mainly on its own ranch for food, following the example of Hispano self-sufficiency, and it touted this autonomous, efficient organization and attendant farm diet to anxious city-dwelling parents. The subsistence farming that for New Mexicans was meant for survival and to perpetuate ancestral traditions was, for the school boys, a way to reinforce their physical and mental abilities.

Only one Hispano and one Native American student ever attended LARS in its twenty-six-year history, but the school was useful to surrounding populations for other reasons. The Hispano homesteaders could sell their crops there and sometimes find employment. The school hired people to look after the livestock and gardens and to maintain the water supply, sewer, and power plant. They also employed a night watchman, an electrician, a plumber, an auto mechanic, cooks, landscapers, dairymen, bookkeepers, several general laborers, and, as buildings were added, carpenters and stonemasons.[84] It became a partnership between the school and the local families from the Hispano villages, the homesteaders, and the few workers who came from nearby pueblos. Interviewee Rubén Montoya, son of Adolfo Montoya, homesteaded on the plateau, supplied meat to the school, and worked as a gardener. He was also the first Hispano student to attend LARS on scholarship.

The Gonzales family worked for the school, too. The wives helped with delivering babies and their children "were assigned various chores at home and for the Ranch School."[85] One of them, Raymond Bences Gonzales, wrote *A Boy on the Hill* to commemorate his childhood on the plateau in the 1930s. This little book is a collection of childhood anecdotes, pranks, games, and mischief that give a humorous, joyful account of life on the mesa before it became a secret atomic laboratory. Gonzales's father was the manager of the trading post and the post office for the school. Raymond Bences recollects, "We all knew each other and lived like one big, happy family, many of us related."[86] He describes the type of work they, as children, were asked to do on the ranch: milking cows, taking care of horses, raising and selling poultry, hunting, and fishing.

In several instances, Gonzales mentions the ties and cohesion among the different communities. His cousin Amador, who was an assistant at the school for employees, and his brother, who taught at the San Ildefonso Pueblo School, organized softball and basketball teams to play games at Los

Alamos and at the Pueblo. His reminiscences portray a cheerful and carefree atmosphere in a microcosm of close families who relied on mutual help:

> I have wonderful memories of the good times I had growing up in Los Alamos. The people that worked for the ranch school were all local people and the families were all very close, helping each other out and sharing goods with each other. It was a quiet and peaceful community, and everyone who lived here was very important to the school because they helped to make it a success.[87]

Gonzales's tale of an idyllic, secluded world up in the mountains counterbalances the asperities of life on the plateau as described previously. Behind Pond's and Brook's romantic vision of living close to nature, there remained the realities of a difficult climate and degrading economic circumstances. The school was under financial pressure, the Anglo homesteaders struggled despite their access to technology, and the Hispano population was affected by the general impoverishment in the valley. Their survival was the fruit of constant hard work and community cohesion, which enabled the "boy on the Hill" to fondly remember this period. Raymond Gonzales's father, Bences Gonzales, joined the Manhattan Project in the war years, working at the post commissary. The family stayed in Los Alamos and Raymond was one of the early graduates of the Los Alamos High School in 1945 before also being hired at the Lab to do technical jobs.

Another person is remembered in Gonzales's book and has come to be identified as a major figure of Los Alamos's history. Edith Warner had settled at the boxcar railroad station of the Denver & Rio Grande "Chili Line" next to Otowi Bridge and in 1928 Connell gave her the job of looking after freight. She later transformed the little house into a tearoom. In her manuscript, she called Otowi "a bridge between two worlds," between the Hill and the San Ildefonso Pueblo.[88] Warner befriended Peggy Pond Church, the daughter of the founder of LARS, who referred to her friend's house as standing "in the shadow of Los Alamos, the mushrooming shadow of violent change in which all of us now must go on living."[89]

The time Warner spent at Otowi and with her friends from the Pueblo (María and Julián Martínez, the famous painters and potters, and Atilano Montoya, who would later live with her) made her aware of that environment's fragility. She wrote in her journal in 1933, "What we do anywhere matters but especially here. . . . Mesas and mountains, rivers and trees, winds and rains are as sensitive to the actions and thought of humans as we are to their forces."[90] Church similarly pointed out the irony behind the selection of Los Alamos as a site to build a weapon of unprecedented destruction; she writes, "How strange it seemed that the bomb which had created such waste and such suffering had been made on the plateau where the ancient people for so long invoked their gods in beauty."[91] Edith Warner died of leukemia at age fifty-seven in 1951 in her new house at the San Ildefonso Pueblo, where she was buried. Although some have wondered whether Warner's disease was radiation induced and if she was "an early downwind victim of Los Alamos testing," there is no information available on the cause of her disease.[92]

Harold H. Brook, Edward Fuller, Alexander C. Ross, Ashley Pond, Albert J. Connell, and Edith Warner, who came to be permanently associated with the Manhattan Project, were all representatives of the Anglo spirit of entrepreneurship, pioneering, and endurance in the West. They were expatriates who would provide a romanticized outlook upon these events. Meanwhile, the Hispano homesteaders and the other local people who were part of the microcosm on the plateau were lost in the meanders of history, barely mentioned in the official narratives of the gigantic undertaking. If their existence came up, it was as a group of anonymous people, not as individuals. When their names did appear in the registers of the Manhattan Engineering District (MED), some of them were misspelled. All were integrated in an immense military and scientific machine that prioritized the success of the mission and relied on the patriotism and sacrifices of locals to support the war effort. The magnitude of the enterprise at Los Alamos—winning the war—at first, seemed to justify the crushing of local histories. But the intricacy of interplay between those local histories and world history lends a complexity to analyzing New Mexico's nuclear story. The plateau connected one of the remotest places in America to world events, but local populations were kept in the dark about the consequences of what happened there.

Three

The Skeleton of a Domestic Nuclear Empire

The transformation of New Mexico is a story of shaping an environment to meet a nation's needs and a story of local people adapting to a socio-economic order established by the authority of government and the influx of newcomers. The circumstances of war permitted the federal government to build a nuclear American West and further its legacy of military and economic control over the region. During World War II and then the Cold War, the relatively barren lands of the western states were perfect for military bases, bombing ranges, test sites, silos, underground control centers, and storage sites. This created a militarized landscape. Military industry was granted locales, as well as governmental protection and support, and was guaranteed enormous profits, which led to a boom in defense-related jobs throughout the West. In the collective conscience, the West previously associated with deserts, cowboys, and Indians has now also become associated with countdowns, radioactive logos, and mushroom clouds. In *The Atomic West*, John M. Findlay and Bruce Hevly explain that whether it was the US Army, the Atomic Energy Commission (AEC), the Energy Research and Development Administration (ERDA), or the Department of Energy (DOE), organizations and agencies in charge of nuclear weapons chose the West to establish centers to pursue their atomic endeavors, "especially the dirtier ones," because they considered the region's undeveloped lands as a possible buffer between people and "the dangers associated with making and testing weapons and storing hazardous wastes."[1]

Militarizing the Landscape

The militarization of the West is rooted in its history of internal colonialism. To be valued by the nation, the West had to be exploitable, and from the start the area proved most useful as a national reserve for commodities and space. In that sense, western territories—later states—were ideal colonies. They had economic use, and their inhabitants, especially in the case of New Mexico, did not have enough political sway to successfully oppose colonial practices. In the meantime, tourists would travel west as if it were a foreign country that was not completely estranged; they would be in search of exoticism while staying within comfortable boundaries. In 1949, many easterners went to New Mexico still thinking it was part of Old Mexico and were reported saying to a storekeeper in Santa Fe: "Be sure to give me change in American money. I don't want Mexican money"; or "When do you hold the bullfights here?"[2] Proof of the permanence of such stereotypes is the long-lived "One of our 50 is missing" column in the *New Mexico Magazine*.[3]

Historians of the American West have long excluded the region from comparisons with imperialist or colonialist nations because of the country's own colonial past within the British Empire. This exclusion left out important aspects of its history that I am trying to rectify here, following the lead of academic works in the late twentieth century that have included the era of American westward expansion in the field of colonial and postcolonial studies "within a paradigm that sees the United States for what it was—an imperial, colonizing state that incorporated the western half of its present-day territory under some rather unequal terms of entry."[4] In the nuclear age, the major difference between the United States and Great Britain or France was that when the latter used, for evident practical reasons, their actual colonial empires overseas—a dominion (Australia) in the British case—to develop their nuclear power, the former primarily used people and space within its own borders because its "national and its imperial boundaries are the same."[5] In both instances, outsiders adopted imperial attitudes in their treatment of the "colony."[6]

Bernard DeVoto introduced the idea of an American "internal domestic empire" that contributed to the shaping of the United States as an exception and as an industrial power.[7] I argue that one can speak of a *nuclear* conquest of the West and of a *nuclear* domestic empire. This new empire, which was created on the foundations laid by the states' territorial history, fits the concept of the unique internal domestic imperialism of the United States. Although the motives and circumstances differed, the creation of America's nuclear weapons complex did follow a blueprint set by nineteenth-century expansionism. By World War II, even though the frontier had been "closed"[8] for fifty years, the West's popular representation as a place where great successes and failures could happen had not diminished. The region, which had been an appealing vessel for national mythologizing as the "land of opportunities," remained a predilection choice for individual and collective enterprises with high expectations and equally high risks—such as "the greatest scientific gamble of history."[9] Many believers in scientific progress of the 1940s thus saw their work as pioneering on the frontier of science and transforming inhospitable lands into thriving communities, which aligned with the purpose of the atomic enterprise that centered on man's desire to conquer natural elements.

Both eastern newcomers and westerners themselves would eventually see scientific programs as a way to wake up their so-called sleepy old communities and give them a chance to step into modernity, to rejuvenate themselves.[10] In December 1945, journalist Hart S. Horn romantically described the village of Organ set just west of the new Alamogordo military reservation as "the original sleeping beauty of southern New Mexico," whose "prosperity graph has had more ups and downs than its mountain peaks." He added, "Although Organ is still a bit doubtful about its reality, a great Army camp has sprung up across the Pass on the wide mesas known to local flower-lovers as the poppy fields."[11] Just as new towns had sprung up on the sides of the railroad tracks in the late nineteenth century, weapons-related facilities would spring up in the wake of the atomic bomb, drawing the map of what would become the nuclear weapons complex. Westerners had already internalized the idea that the West was a place where anything was possible, including the sudden materialization of an Army camp. They had been advertising this trait to attract individuals in search of these western promises. When soldiers arrived at the small Alamogordo railroad station to reach the military base there in the 1940s, the first sight they had was that of the town's sign across the station: "Alamogordo. New Mexico. Little City of Big Trees and Opportunities."[12]

My purpose in using the colonial framework is to confront western mythologies and the mindset of colonial rulers to the experiences of local residents. On the one hand, the arrival of nuclear science was perceived as a rupture in the state's history by local populations who first saw science and progress, the spearhead of this singular conquest, as their way to a more prosperous and autonomous society—until the rise of activism in the 1970s. On the other hand, the actions of the Army, the federal government, and the scientists hinged on these mythologies and were influenced by this mindset, thus creating continuity. Patricia Nelson Limerick, who the coined the expression "Atomic West,"[13] identified certain enduring characteristics of the West: the remarkable presence of the federal government, the boom and bust economies often in the extraction field, the "mythologizing of the West as a place of romantic escape and adventure," and its use as a dumping ground.[14] These elements form the pillars of New Mexico's scientific conquest: the role of the government, the ups and downs in the uranium industry, the influence of western myths over the scientists' experiences, and the creation of a nuclear waste repository.

Changes in the West "came in manner very much shaped by the western past" and "flowed through familiar channels," according to Richard White.[15] The first of these "familiar channels" was the military. In the eighteenth and nineteenth centuries, the West served as a protective barrier against foreign powers and Native Americans. It was dotted with military forts. By the beginning of the twentieth century, these had been deserted, but the area would renew its connection with the military with the Manhattan Project. An article in June 1941 announced, "A new use has been found for New Mexico lands—at least, those areas heretofore thought to be useful only for grazing of sheep or goats, which thrive where even a longhorn cow would starve. The Army flying corps has discovered that such lands—waste or practically that—are admirably suited for bombing practice."[16] As shown in this comment, military planners saw the West's old liabilities as virtues. The region's vastness, low population density, isolation, and arid climate that had been detrimental to economic development now were assets for the location of military bases.

From 1941 to 1945, the War Department located twenty-one separate military bases, training centers, and prisoner-of-war and Japanese internment camps in New Mexico.[17] Some of these were eventually integrated into the local scientific network. They became centers for research and the development of technology, so the new economy merged the three components: science, the Army, and corporate industry. The military landscape created precedent for land acquisition and considerably eased the atomic enterprise. The militarization of New Mexico manifested in the establishment of four Air Force bases (AFBs) and one Army base.

The Walker AFB near Roswell was created as an Army Air Corps flying school in 1941. In the postwar years, it became the Roswell Army Air Field and then Walker AFB. This base closed in 1967 for financial reasons when expenses on the Vietnam War were at their highest. Its units were relocated from New Mexico to Alaska or New York, and, since then, the facility has been known as the Roswell International Air Center, where aircraft is stored, and where the Eastern New Mexico University built a campus. The remainder of the base land and buildings that have not been reused were left abandoned. This abandonment illustrates how much military installations are dependent on federal budgets and can thus disappear as fast as they appear in time of war, taking with them the socioeconomic activity they were generating in the area.

The other three bases are still active and have become substantial contributors to the local economy. The Cannon AFB near Clovis was created in 1942 on the site of the Clovis Municipal Airport. The Clovis Army Airfield—which was renamed Cannon Army Airfield in 1957 to honor General John K. Cannon—was used as a training ground for heavy bomber pilots during the war. In late 1942, the taking of lands for the establishment of an aerial gunnery range for the Clovis Air School was decried in a letter to Governor John J. Dempsey (1943–1947). The attorneys who authored the letter warned Dempsey that they took the matter seriously "as the area takes in what is probably the best grazing land in New Mexico."[18] A protest was signed by "458 citizens owning 631,300 acres of land, 31,000 head of cattle from which was produced practically ten million pounds of beef in the year 1942." The Farmers & Stockmens Bank also complained to the governor that "with all the waste land there is in New Mexico it seems absurd to us for the government to select one of the best cattle raising districts within the state to take from the food production which is so badly needed at this time."[19] This comment is striking because it reveals that some residents were equally influenced by the preconceptions of emptiness, believing that if the area where they had settled did not fit the description, another neighboring area would.

This proves that militarization did not go without resistance especially when it threatened agriculture, a pillar of the state's economy. In Clovis, the project of a bombing range was abandoned but, further south, the establishment of the huge White Sands Missile Range (WSMR) was successful perhaps because the area was considered "wastelands," despite the resistance of a scarce but productive rural population. After the war, however, when the government was closing bases, members of the Melrose Rotary Club, twenty miles from the base, sent a petition to US Secretary of Agriculture Clinton P. Anderson (1945–1948), who was from New Mexico, to ask that he use his "influence in trying to get this base left here as a permanent fixture in New Mexico." They believed its continuance was "for the best interest of all concerned."[20] New Mexicans would soon grow accustomed to asking their representatives to use their influence to maintain military and government facilities that generated economic activity in the state.

In Albuquerque, the history of Kirtland AFB began in the late 1920s and is intertwined with the early days of aviation: the qualities that made Albuquerque an aerial hub were similar to those given by the Army for selecting sites in the state. In 1927, Charles S. Lindbergh had flown across the Atlantic; as a result, Americans became "air minded."[21] Following Lindbergh's success, Frank G. Speakman, an employee of the Santa Fe Railroad who homesteaded a ranch at the foot of the Manzano Mountains, became interested in aviation. He and his partner, William L. Franklin, leased 140 acres of flat land on the East Mesa in Albuquerque to build an airport in 1928. Then James G. Oxnard, a New York air transportation promoter, bought out Franklin's share and established Aircraft Holdings. The airfield on East Mesa became known as Oxnard Field and welcomed national celebrities during the 1930s golden age of air racing and record setting.

As early as 1939, the Army became interested in Albuquerque to establish a training base for aircrews. It leased two thousand acres from the city on the East Mesa, close to Oxnard Field.[22] The Army set up a bombardier school, which was designated as the Air Forces Advanced Flying School on this Albuquerque air base. In 1942, the base was renamed Kirtland Field in honor of Colonel Roy C. Kirtland, one of the nation's pioneer Army aviators. That same year the Army began condemnation proceedings, offering $80,000 to Airport Holdings. However, Frank Speakman and James Oxnard objected to the amount and eventually obtained $95,000.[23] This example demonstrates that, given the opportunity, landowners could raise the price of their compensations. This would not be the case, however, for other New Mexicans who lost land to the Army during the war and in later years. The bombardier school was the base's chief activity, but, as early as March 1945, this acquisition proved useful to the Los Alamos scientists who needed to make sure the bomb would be a viable airborne weapon. The military property subsequently became an extension of the Los Alamos Laboratory when its Z division was moved to Albuquerque in 1947. Speakman's aviation dream became the locale of what is now the Sandia National Laboratories.

The last Air Force Base, which was also created in 1942, is Holloman AFB, west of Alamogordo in the Tularosa Basin. First named the Alamogordo Army Air Field, the base was established as training grounds for aircrews and was renamed in 1948 to honor Colonel George V. Holloman, a pioneer in guided missile research.[24] Initially, the range was intended as an extension of the airfield, but missile research and testing became its prime objectives in the postwar years because of its proximity to the area now called White Sands Missile Range (WSMR). As the weapons became bigger, more land was needed to test them. The history of the White Sands military reserve is a symbol, along with that of the Los Alamos Homesteaders, of how the Army imposed its will on local people, expropriating their property in the name of the war effort and national security. The area first known as the Alamogordo Bombing and Gunnery Range (1942), then the White Sands Proving Grounds (1945), and finally White Sands Missile Range (1958) is the largest overland military test range in the country (3,200 square miles). It is interesting that the second-largest US military reservation, Fort Bliss Military Reservation, created in the 1850s, also spreads over southern New Mexico in Otero County, but its headquarters are in El Paso, Texas. Fort Bliss has primarily impacted the economy of El Paso even though 80 percent of its lands are situated in southern New Mexico. The presence of this military land dating back to the period of westward expansion can be interpreted as the physical representation of one of White's "familiar channels."

Sacrifice Land for the War Effort

WSMR is in fact a combination of various lands that were handed over to the Army: Fort Bliss Antiaircraft Firing Range, Doña Ana Target Range, Castner Target Range, and Alamogordo Army Field's Alamogordo Bombing Range.[25] In 1942, the first step toward its creation was the acquisition of both public and private lands to establish the Alamogordo Bombing Range. The Army was interested in nearly 1.3 million acres out of which 21 percent were state owned but contained sections under lease for grazing, 2 percent were in private hands, and the rest was public domain.[26]

One may note (as a 1942 article in the *New Mexico Magazine* did when describing the area) that some of the ranch holdings coveted by the Army had been owned by famous politicians—New Mexico senator and secretary of the interior Albert Bacon Fall (1921–1923) and his successor as senator, Holm O. Bursum (1921–1924).[27] The ranch Fall owned near the northwest corner of the Mescalero Apache Reservation was called the Three Rivers Ranch. Between 1912 and 1923, Fall expanded his holdings by a factor of ten and made repeated efforts to take Mescalero land for a national park. As for Bursum, he was first elected sheriff of Socorro County in 1894 before becoming chairman of the Republican Party Territorial Central Committee in New Mexico from 1904 to 1911. Over this period, he was instrumental to New Mexico's access to statehood. Then, in 1921, he succeeded Fall in the Senate. Bursum and Fall—now secretary of the interior—worked together to revive "national park proposals, opening of Indian lands to mineral exploration, quieting title to Pueblo Indian lands contested by non-Indian owners, and easing of federal restrictions on western resource development."[28] Both politicians, therefore, were active in the "colonial" past of New Mexico, particularly through their attempts to quiet Indian land claims with the Bursum bill that would have legalized the claims of non-Indians to Pueblo lands they had acquired before 1902 and enabled state courts, which rarely sided with Indians, to have jurisdiction over land disputes and water rights. The Pueblo Indians united against the bill. With the help, among others, of Commissioner of Indian Affairs John Collier (1933–1945), they put pressure on Congress until it was defeated. In 1924 Congress passed the Pueblo Lands Act, which recognized the land rights of the Pueblo peoples.[29]

During the acquisition proceedings for the bombing range, the New

Mexico Land Office "used its influence and good offices in helping it to make satisfactory arrangements for the leasing of privately owned or leased parcels."[30] The Army generally was able to rely on the state's support; but when it came to taking such a vast portion of land in the Tularosa Basin, the negotiations between New Mexico Governor Dempsey and Secretary of War Henry Stimson (1940–1945) show that the state had reservations about giving up some of its best grazing lands. Since the mid-1800s, the Basin had been a traditional grazing area for Hispano and Anglo New Mexicans, particularly immigrants from Texas. The open range system had disappeared with the new public land laws and the spread of the cash-based economy, yet ranching remained a very important source of income in these counties.

On January 8, 1943, Stimson wrote to Dempsey about a possible land acquisition by the War Department for temporary and permanent installations in areas embracing grazing rights and privileges, including the Alamogordo Bombing Range and the Roswell, Hobbs, Carlsbad, and Kirtland field practice bombing ranges. According to War Department policy, the exercise of federal and state grazing privileges had to be suspended. But in the case of rights derived through a lease from the state, consent of the lessor was required, contrary to leases in the public domain: "This is accomplished by purchasing from the lessee a sub-lease for a specified period of years, containing a provision suspending the exercise of the grazing privileges *for the period of the lease and any renewal thereof.*"[31] This regulation implies that the taking of land was initially meant to be temporary and that grazing rights and privileges would be resumed after termination of the Lease and Suspension Agreement. The document also required that "the Grantor [typically a rancher who leased grazing land from the state] agrees to continue paying the taxes, State or other lease fees, other overhead costs, and permit fees unless payment of said fees is cancelled or forgiven by the Government."[32] According to this section, landowners not only had to relinquish their rights to the land but also had to continue paying taxes on the land they could no longer use. This was the case until the government purchased all the land in the 1970s. The moment the grantor signed the agreement, he granted "to the United States the right of immediate occupancy" and use of the land "for any purpose whatsoever." In cases where land was needed for permanent installations such as "airfields, auxiliary landing fields and ordnance storage areas" that would "completely destroy its value for grazing purposes," the government would acquire a fee simple title.

In his answer later that month, Governor Dempsey strongly emphasized the importance of these lands not only for the New Mexican economy but also for its tax base as institutional lands contributed to the financing of education, charitable, and penal institutions.[33] One map, probably made in the early 1940s and donated to the WSMR Museum by the Henderson family, illustrates the ranch units in the area of Salinas Peak in the upper half of the range. Each unit is represented with both the patented and the leased grazing lands. State "school" sections are also visible. These were trust lands given to New Mexico by Congress under the Ferguson Act of 1898 and the Enabling Act of 1910, and they are part of the sections Dempsey referred to in his letter. The grazing rights ranchers paid to use school sections went to the New Mexico State Land Office to benefit public schools in the state. The governor estimated that the transaction was going well as it had been approved by the commissioner of public lands that there should be no difficulty in rehabilitating the "livestock men in this area" after the war.[34] Nonetheless, anxiety over how the disappearance of ranches and farms would affect the local economy were expressed in a letter by Lewis N. Gillis, president of the Alamogordo Chamber of Commerce, to the El Paso Chamber of Commerce in August 1943:

> The establishment of this Range will seriously and adversely affect
> Alamogordo and Otero County and we think, deprive El Paso business
> houses of a great deal of business which they have heretofore enjoyed
> from this area. . . . We do not want to impede the war effort in any
> way, but after a careful survey, we are convinced that the establishment
> of this Range in this County will require a sacrifice on the part of our
> citizens far beyond the necessity of or value to the war effort.[35]

In response, Chris P. Fox, vice president of the El Paso Chamber of Commerce, voiced his sympathy, "I know that the Congressional delegation from your state have all expressed a desire to try and avoid a usage of this land, under the conditions outlined, or until it has been fully established that there is no other way out."[36] In the end, the negotiations did not go in the direction that Gillis desired.

In the meantime, Commissioner of Public Lands H. R. Rodgers went to Washington on the ranchers' behalf to ask for fair compensation. He declared upon his return that the War Department was sympathetic to their situation and gave assurance that "every consideration would be given to the problem and to the welfare of the people involved."[37] The following May, an article in *New Mexico Magazine* emphasized, "There is no idea of standing in the way of the project merely that of trying to prevail upon Uncle Sam not to dislodge the cattlemen, sheepmen, farmers, and others now located on the site with the minimum of inconvenience for the conduct of their business and enabling them to continue it in the interest of the war effort without undue interruption."[38] Rodgers's idea was to make an exchange that would replace the portion of grazing lands owned by the state by some other lands elsewhere and seek "suitable remuneration for the private owners." Nothing could be done for the individuals who had grazing permits on federally owned land that were canceled by the Department of the Interior, but Rodgers entered negotiations with the Departments of War and of Justice on the question of state permits to "give all interested parties an opportunity to state their case before taking legal action."[39] Like Lewis Gillis, Rodgers took action because he was driven by anxiety for the economies of the Otero, Lincoln, Doña Ana, and Socorro Counties that were to "suffer considerably in the loss of personal property from their tax rolls, in livestock, and also in improvements."[40]

The governor's correspondence and the press articles make it obvious that the State of New Mexico, although willing to participate in the war effort, was reluctant to relinquish some of its best economic assets that fueled its fragile economy. Negotiations clearly stated that the lands would be returned after the war, yet the expropriated ranchers never recovered their properties. Concerns did not end there; new issues regarding the range continued to arise. For example, new telegrams were sent in April 1945 to the governor by the chairman of Doña Ana County Commissioners and the president of the Las Cruces Lions Club to inform Dempsey of the protest against closing Highway 70 because of the rocket range as a detriment to their town.[41] Then, in the 1950s, the White Sands Proving Ground extended further east toward the Sacramento Mountains, dispossessing around fifty more ranches. Protest reached the state and federal levels in politics and in court, but the most visible protest came from the individual level.

After World War II, some of the ranchers of the Tularosa Basin who experienced the loss of their lands, houses, livelihoods, and lifestyles entered a legal battle against the US Army and the federal government. Their voices acknowledged one crucial fact: the area now known as WSMR was not wasteland; it was not empty. A section of the WSMR Museum dedicated to the ranching era in the Tularosa Basin explains to visitors that, for over half a century, the Basin was home to more than a hundred sheep, cattle, and goat ranches. The San Augustin Ranch, for example, formed much of the land that is now military owned. This ranch, operated by the Cox family since the 1800s, spread over more than 150,000 acres; in 1945 the Army used eminent domain to take more than 90 percent of it, and the family now lives in the remaining land and house just a few miles west of the main post housing area.

Most ranching families, like the Coxes, had inhabited the area since around 1880. They lived up to twelve miles from any neighbors and sometimes as far as fifty miles from a town. The WSMR Museum exhibit describes the ranchers' lives using the romantic prism dear to the western literary genre, the stereotyped nostalgia often present in the United States when referring to the country's agrarian past, and the lure of freedom through agricultural activities that had sustained the recreational home-steading movement:

> If they got in the mood to visit, they saddled up their favorite horses and took off. Sometimes, entire families would ride twenty miles to attend a dance . . . When they were children, the canyons and ridges, rocks and trees became the landmarks of their neighborhood. As adults, they scratched from the very earth the materials—particularly rock and adobe—with which they built their homes. They lived from the land and marveled in its beauty. . . . The things modern women take for granted—hot and cold running water, washing machines, microwaves, hair dryers—had either not been invented or had not found their way into the remote fastnesses [sic] of Deadman's Canyon and Lava Gap. If families had electricity at all, it usually was generated by a wind charger, stored in car batteries and used to power lights and radios.[42]

The role of women is particularly emphasized to illustrate the difficulties of life in the desert and to celebrate their resourcefulness. The women engaged in various activities that were typical of ranching, for example, cooking beans, baking sourdough, canning goods, making jerky, and frying beef cut from carcasses on the porch during cold weather. Giving birth took place at the ranch without any doctor or nurse, just accompanied by a relative. In the 1940s, ranching in the area had not much changed from the 1880s. Ranch life was similar in many ways to village life in New Mexico's native communities, including self-sustenance, seclusion, and solidarity. Families owned gardens to grow fruit and vegetables and kept a cow to produce dairy goods. Flour, sugar, and other staples were bought in town. Ground water was only used to water livestock because it contained minerals. The ranchers drank water collected from the metal roofs in cisterns.[43]

The exhibit panel ends with a paragraph on the "supreme sacrifice" that these ranchers made when they gave up their land to aid the war. Lewis Gillis had used the same words—"sacrifice" and "war effort"—in his telegram. The word "sacrifice" is important as it etymologically means to offer something, especially a *life*, to a deity. In this case, one may say a *way of life*. The museum insists on the ranchers' patriotism: "Not only did they give up their homes and livelihoods in support of the war effort, but many sons and husbands also served in the military during World War II. Rancher Eda Anderson Baird reached the rank of lieutenant in the Army medical corps."[44] The exhibit, while providing interesting insights into the ranching era in the Basin, gives a romanticized view of the ranchers' lives and of their reluctant renunciation of the ranching lifestyle.

Not all testimonies, however, recount the event from a patriotic angle. In 1942, when the first range was created, about fifteen ranches were located on the million acres coveted by the Army, including one owned by Tom McDonald and his sons. In the 1955 extension, fifty more were targeted among which was John Prather's ranch. Both McDonald and Prather became nationally known for their determined resistance to the government and the Army.

The McDonald family, originally from Ireland, had moved to New Mexico from Texas. Mike McDonald first homesteaded in the Mockingbird Gap in the La Luz area in the late 1870s. Tom McDonald went further south. All the McDonald children of the pre-1942 era either worked on or owned ranches. When they were not looking after cattle, the children would attend school in Oscura, about thirty-five miles away. David McDonald recalled that, as a child, his first saddle had been a longed-for Christmas gift. Like in other New Mexican communities, interfamily help was essential to survive, contrary to what the iconic imagery of the West and the lonesome cowboy mystique would portray.

Four McDonald ranches were taken by the Army to build the bombing range. The family did not know that the Army was interested in their properties until someone came to evaluate the land, so they had very little time to vacate. They left equipment behind, believing it would be there when they returned; but things were stolen or allowed to deteriorate, and their cattle was driven to pasture or trucked out for selling. David McDonald complained that the New Mexico practice of valuing ranches based on patented land and related grazing leases was not honored by the government: he speculated that the way compensations were determined may have been arbitrary or that some ranchers were able to use their political influence, bribery, and better legal representation to raise them. The examples of Oxnard Field and Clovis mentioned earlier would tend to confirm this statement. Dave McDonald and his brother Ross went to court and managed to slightly increase the amount, but their legal battle lasted for several years.

David McDonald said they understood that their land was needed during the war, but they were anxious about the loss of their livelihood. The amount of the compensations they received did not enable them to reproduce their lifestyle elsewhere. The families were not given any funds for lodging or food during the move. Because cattle prices dropped, selling their animals did not help. He further contended that the ranchers were devastated by the taking of their ranches; and, in addition to the family members they had lost at war, they lost their homes and the trust they had in the federal government. This was clearly expressed by Helen Wrye, another rancher in the area and witness of the Trinity Test. She talked wistfully of a time when "people weren't afraid of the government." "It was a time of innocence. People were trusting. We had never heard of an atomic bomb," she said in 1995.[45] David McDonald voiced his bitterness at being treated as if his land had been "free land." He thus became a victim of one western myth while being the representation of another: the cowboy on the range, who, incidentally, had also taken possession of "free" land.[46]

John Prather came to New Mexico with his brother Owen from Van Zandt County, Texas, in 1883. The two men were also seeking "free" profitable land at the very end of the westward migrations. They arrived in Lincoln County, which encompassed much of the Tularosa Basin at the time. John settled below the Sacramento Mountains to raise cattle, and Owen developed a sheep ranch. Interviewee Irving Porter was eleven years old when he was taken in by the Prather family in Piñon. He described John's ranch as heaven, "an oasis in the middle of the desert." The family first ranched on the Agua Chiquita near Weed in the Lincoln Forest before going further west to the "flats." John started with a homestead on a school section and eventually became known as the "Mule King." To enlarge his property, Prather purchased the land from ranchers in the area who homesteaded only one section of land, which often was not enough to make a living. This acquisition process of smaller parcels, and thus the creation of large ranches, was one consequence of the private property system. Without community lands to help smaller landowners make up for limited resources on their properties, they could not build a viable livelihood and were eventually absorbed by their neighbors. The western genre fueled on such stories of competition and rivalry, adding another layer to the cowboy mystique as a fierce guardian of his property. Although both McDonald and Prather would live up to that reputation in their resistance to the Army, the ranchers' testimonies tend to paint a picture of neighborly cooperation rather than hostility. At the time of the government land grab in 1955 to extend the White Sands Proving Ground eastward, Prather was eighty-two years old and owned eight sections (5,120 acres) of patented land that the Army wanted. It was the largest ranch in the area.[47]

Another ranching family in the area who had deep roots in the state was that of Ernest Aguayo's. His grandfather—a lawyer, schoolteacher, and justice of the peace in Lincoln County—witnessed the Lincoln County War, one of the most epic events in New Mexican history, and befriended Billy the Kid. This heritage of the Old West has often been used to introduce the military reservation. One article read, "The country of Eugene Manlove Rhodes, of which he wrote so picturesquely and romantically, and of Billy the Kid, which is still 'cow country,' soon is to resound again to the sound of crackling gunfire—gunfire of greater volume than Gene Rhodes or the deadly Kid ever dreamed of."[48] The violent and mythical past of the region seemed to make New Mexico a perfect match for heavy weaponry.

In the case of Aguayo's family, the western heritage was mixed with a Spanish heritage reflected in the Castilian language spoken by Aguayo's father and by the name Aguayo, which refers to a river where the prince of Spain was saved by three brothers. Ernest grew up on the Vega Ranch on the Carrizozo Flats below Nogal before the family moved to the Tortalito Ranch further to the southwest. His father was the first person to take out a permit to lease land for grazing from the Lincoln National Forest in 1913. Once he finished his schooling in Nogal and Carrizozo, Aguayo became a cowboy and worked with his father and his brother until his father retired and sold the ranch. Aguayo later acquired his own land on the east side of the Oscura Mountains; he owned 640 acres of patented land and a grazing lease for eighty-four cattle and saddle horses. He complemented the ranching activities by working for the US Forest Service and with the Southern Pacific Railroad. This occupational evolution reflects the regional trend toward wage work as a way to supplement income from declining agricultural activities. The ranch was taken by the government in 1941. Aguayo received $1,000 in land use fees but was given no chance to remove any of his belongings. He estimated that it took three months from the time he was notified about the taking until he had a chance to meet with the government attorneys. Aguayo had every intention of going back and starting over after the war, but the government takeover ended his project. In the postwar economy, he turned to other occupations: he worked as a blacksmith in Mesilla Park and later for the Empire Zinc exploration division, and, finally, went back to work for Southern Pacific Railroad. He later took an active part in the protest and legal strife that was undertaken by the WSMR Ranchers.[49]

Aguayo's, Prather's, and McDonald's resistance attracted media coverage and public sympathy to the ranchers' cause but also manifested the disappearance of a lifestyle. The ranchers' plight is a point of convergence between western mythologies and realities. Some people, such as the Prathers, arrived in Lincoln County looking for lands free for the taking and aspiring to the rugged life of westward migrant farmers only to be stripped later of that land through a mechanism largely influenced by concepts shaped in the westward migration era. As in the selection of the isolated Pajarito Plateau in Los Alamos, there is irony in the ranchers' situation. They received the land from the state, which had initially acquired it through violence against the region's ancestral inhabitants, and developed it, readily observing the open range myths; and then, the federal government took the land as its own from them just as easily.

Local and World Histories Meet

By 1942 the work on the design and building of an atomic bomb was being conducted in university labs throughout the country.[50] The creation of so-called Project Y was meant to accelerate scientific progress on the conception of the bomb by centralizing research on components that had been geographically separated and compartmentalized for security reasons. In "appealing directly to [General] Groves' obsession with security, [J. Robert] Oppenheimer proposed that all the research be consolidated in a single laboratory, located in an *isolated* region, where the scientists could converse freely among themselves and secrecy could still be maintained."[51] Oppenheimer and Arthur Compton first considered settling the lab at the Clinton Engineer Works in Tennessee or in Chicago, but neither location was isolated enough.

Groves favored remote areas where "nearby communities would not be adversely affected by any unforeseen results from [their] activities."[52] In other terms, the people living near the selected site should not be prone to conflict in case something went wrong. Clearly, isolation was the key criterion in the siting process. Not without irony, the same elements that had made New Mexico a foreign, distant, almost unwanted region suddenly became fundamental assets. Land grabbing entered a new phase with the atomic age. Now the federal government needed "empty," cheap land for its atomic and nuclear pursuits, and government officials generally perceived the West as just that: a reserve of wastelands that would finally have their use. What the project officials desired was a place on the frontier, far from Anglo civilization. Yet, like in the Tularosa Basin, the microcosmic world on the Pajarito Plateau was hardly construed as a frontier by those families who had worked out a complex interdependent system to maintain life in this unforgiving environment.

Although the region's colonial past was a large determinant in the arrival and the spread of the scientific conquest, it was nonetheless a violent rupture in an already inhabited, organized world. Only from the perspective of conquerors does the notion of an empty place on the edge of civilization exist. The Hispano homesteaders were "barely noted" by the Army official when he assessed the locale even though they would be the majority of expatriated residents at Los Alamos. Peter Hales notes, "Because they were not 'authentic,' in nostalgic Anglo terms—that is, not reservation-dwellers, pottery-making Indians, but small-time capitalists, cultural half-breeds who eked out their livings on marginal lands—these residents of the mesa hardly existed in the Western mythology invented by the émigrés who surrounded them."[53] They did not correspond to the preconceptions Americans had about what "true" inhabitants of the West were supposed to be. Even though they were the participants of a genuine local culture, their existence on the fringe of western mythology, which then defined the identity of places such as New Mexico, meant being ignored.

Lieutenant Colonel John H. Dudley was appointed to investigate possible sites. First, the location had to support an estimated population of 265 people. Oppenheimer had imagined that six scientists assisted by a few engineers, technicians, and draftsmen would be enough. Thinking of families, plumbers, electricians, guards, storekeepers, and school teachers, Dudley raised the number to 450; by November 1942, it had reached 600. Second, the proposed laboratory would have to be built at least two hundred miles inland because coastal locations or one abutting foreign borders would pose too many safety threats from enemy attacks. Third, the climate should allow year-round experiments and construction. Fourth, existing housing facilities were needed to enable the first 6 scientists to start work immediately. Fifth, though isolated, the site would have to be accessible with ready transportation to permit cars and trucks to deliver people and material. Last, in this spirit of isolation and security, the topography of the area should be a natural bowl surrounded by hills close by to put up fences and install guards at the top.[54] Other criteria included access to power, fuel, and water, adequate testing grounds, sparse local population, easy acquisition of the land, and a local labor force. The last two would affect northern New Mexicans the most.[55]

A site near Los Angeles, California, was first considered, but Groves thought the area was too populated and would pose a chance that scientists would be tempted to socialize with the faculty at the California Institute of Technology. He also rejected a site near Reno, Nevada, and Oak City, Utah, because of the number of farmers and their families who lived there. The choice of New Mexico has been attributed to Percival C. Keith, a member of the planning board for the government's Office of Scientific Research and Development set up in December 1941. Keith's sons, Percival and Christopher, were summer campers at LARS. He wanted to locate both the laboratory and the uranium enrichment facilities in New Mexico from the start.[56] Jémez Springs, a deep canyon west of Santa Fe, was the first site Dudley selected in the state. When Groves came to visit on November 16, 1942, he immediately rejected it because there were too many farmers to dispossess and the mesas that surrounded the canyon would be an obstacle to expansion. Oppenheimer also thought the cliffs would have a depressing effect on the scientists' morale and the sound of test explosions would probably reverberate against the walls.[57]

Groves, at all costs, wanted to avoid having trouble dispossessing landowners. His main concern with finding locations in the West was to avoid Indian presence, a point he repeatedly mentions in his memoirs: "As we went along the road to the north, we drove by many small Indian farms, and I began to have misgivings about the troubles we would have in dispossessing the owners." In searching for a test site in 1944, he comments, "I added one special prohibition: that it should have no Indian population at all, for I wanted to avoid the impossible problems that would have been created by Secretary of the Interior Harold L. Ickes, who had jurisdiction over the Bureau of Indian Affairs."[58] In the end, the project encountered no difficulties with Native Americans in the early years of LANL but its presence became increasingly problematic for its neighboring Native communities later.

After inspecting Jémez, the small group proceeded to evaluate the Ranch School following Oppenheimer's suggestion. Los Alamos did not meet all criteria. For one, it was no natural bowl, which is why Dudley had initially discarded it. But the bowl concept was inverted: the cliffs around the mesa made the site just as inaccessible and the altitude permitted views important to both Oppenheimer for anti-depressant purposes and to Groves for security reasons.[59] The excellent air and rail facilities in Albuquerque, which would facilitate shipment of material, also tipped the scales toward the choice of Los Alamos. In fact, the Army's Albuquerque Engineering District "was assigned the task of building the laboratory and living quarters for the scientists who would be working on the hush-hush Manhattan Project."[60] The general later noted that the real problems with the locations, which were only revealed a few months into the Project, were the water supply, because controlling the residents' use of water was almost impossible; the road, because it was ill-adapted to heavy vehicles; and the lack of skilled labor in the vicinity.[61] Low-skill labor was plentiful and indispensable, but the habit of viewing the local workforce as unskilled persisted through the years despite manifest progress in educational attainment.

One detail in the selection process that gained momentous importance is that Oppenheimer suggested the site. In 1921, Oppenheimer had stayed at the Los Piños guest ranch run by Katherine Chávez Page in Cowles. On one of their lengthy horse riding excursions, they had gone from the village of Frijoles to the Pajarito Plateau through Valle Grande. That is when he had seen Los Alamos for the first time. Fascination with this happenstance possibly comes from the physicist's special connection to the Land of Enchantment. Oppenheimer had written, "My two great loves are physics and New Mexico, it's a pity they can't be combined."[62] All his summers during the 1930s were spent at a ranch his father rented in the upper Pecos Valley called Perro Caliente (hot dog). His love for the area expressed itself through his sojourns, his career choices (his professorship at UC-Berkeley enabled him to be closer to the desert), but also through his letters and poetry.[63] The height of irony, however, is that this selfish desire led him to irreversibly alter and even destroy the magnetism he loved about the region. That is perhaps why the anecdote became legendary. Much of the growth and spread of nuclear research and development centers, high-technology facilities, and military installations in New Mexico are the result of Oppenheimer's decision to take Groves to visit the LARS.

The figure of Oppenheimer has also fascinated historians and biographers because of his many faces: the romantic, the genius, the hero, the leftist, the activist, and the fallen scientist. His relationship with New Mexico combines many aspects of his complex personality and represents the paradox of many travelers to the state who cherished the region's identity of preserved landscapes and ancient places but, through their actions, inevitably contributed to compromising its beauty. After the war, he expressed his regrets in a letter to a LARS headmaster: "I am responsible for ruining a beautiful place."[64] Edward Teller claimed that Oppenheimer's answer to the question "what should be done with Los Alamos" after the bombing of Hiroshima was "Let's give it back to the Indians."[65]

In the end, it was a conglomeration of chance and determined circumstances that designated New Mexico as the ideal recipient of a scientific revolution and economic renaissance. Not only did the state fulfill almost all the criteria set by the Army in 1942, but it also seemed ripe for yet another invasion, one that would bring opportunities for development.

The consequences of the first American military conquest and the effects of internal colonialism were still taking a toll on the state's economic base as it moved from subsistence agriculture to dependence on far-off markets. Industrialization was slow and virtually absent outside of urban areas; poverty was rampant throughout the state, forcing the population to migrate regularly across its borders to find employment. Yet, many New Mexican rural families such as those who lived on the Pajarito Plateau and in the Tularosa Basin maintained their forbearers' way of life. Traditions, solidarity, and hard work were valued above all; the most grueling experience was the regular separation forced upon families to survive in the expanding cash economy. Larry Torres, Taos educator and historian, remarked on the transition from seclusion to nuclear science and high technology:

> When you consider that the last Moorish stronghold at Granada fell in 1492, and that the Spanish start coming to conquer Mexico, and up to the Rio Grande valley, they are coming straight out of the Middle Ages. They have not undergone the Renaissance, so that the people here are isolated for centuries. Our religion, our traditions, our customs, our way of looking at people are very medieval, still reflected in the language of today, which is very archaic. Taos Spanish is three hundred years out of fashion, so that we have scholars from Spain, who come here to study the language of Cervantes here among us. *What happens when you have a society that goes from this to that with no transition or stages?*[66]

This book attempts to bring an answer to Torres's question and show how the impact of sudden, drastic change on traditional societies is a two-stage process, from exhilaration to disillusionment, for the affected individuals.

Four

The Manifest Destiny of Atomic Scientists

Specialist of Los Alamos's history Jon Hunner argues that the presence of the Army in New Mexico was a tradition dating back to the territorial period and that the Manhattan Project was merely "a new wave of military and industrial influence and expenditure in the region."[1] The difference, though, is that while previous military conquests were generally accompanied by scientific research, it was secondary. In Los Alamos, science was the main purpose and the Army was the means. Land acquisition for the site followed the pattern of military settlements elsewhere in New Mexico but was accomplished much faster and without any interference. The displacement of Plateau inhabitants revealed, even more strikingly than in the Tularosa Basin or on the military base sites, the hierarchy among evicted people. Some were able to negotiate while others were hastily and forcibly removed in the name of national defense. Approximately fifty-four thousand acres of land were deemed necessary to build the complex. One of the main reasons for the selection of the Los Alamos site was the fact much of the land already belonged to the federal government in the form of the Santa Fe National Forest. The local population held grazing permits to land there, but they were transferred to surrounding lands controlled by the Baca Land Company, setting up a ranching equivalent of sharecropping. Out of the 54,000, approximately 8,900 were privately owned, the largest single property belonging to LARS.

The Big Invasion

After the first visit to the school by General Groves, Oppenheimer, and Lieutenant Colonel Dudley in November 1942, the Army started posting guards and sending more people to assess the locale. The schoolboys dubbed this "the Big Invasion."[2] General Groves emphasized in his version of the condemnation proceedings that the headmasters were more than happy to cede the land to the Army due to a staff shortage: "I was most relieved to find that they were anxious to get rid of the school, for they had been experiencing great difficulty in obtaining suitable instructors since America had entered the War, and were happy indeed to sell out to us and close down for the duration—and, as it turned out, forever."[3] Actually the school director, A. J. Connell, who was indeed facing financial difficulty, was prepared to negotiate a sale but only until the end of the war. That he was "anxious to get rid" of it challenges credulity.[4]

Connell refused all the offers from Army real estate men, engineers, scientists, and inspectors until he received a letter from War Secretary Henry Stimson on December 1, 1942, that read:

> You are advised that it has been determined necessary to the interest of the United States in the prosecution of the War that the property of the Los Alamos Ranch School be acquired for military purposes. Therefore, pursuant to existing law, a condemnation proceeding will be instituted in the United States District Court for the District of New Mexico to acquire all of the school's lands and buildings, together with all personal property owned by the school and used in connection with its operation. Although the acquisition of this property is of the utmost importance in the prosecution of the war, it has been determined that it will not be necessary for you to surrender complete possession of the premises until February 8, 1943.[5]

A week later, the faculty, staff, and students were told the news. One student understood what was happening to his school: Sterling Colgate, a senior, had recognized Oppenheimer and Ernest O. Lawrence when they had come to visit the school at the end of November. He recalls, "These two characters showed up, Mr. Smith and Mr. Jones, one wearing a porkpie hat and the other a normal hat, and these two guys went around as if they owned the place." While still attending classes, students saw the trucks and bulldozer begin their work on the "demolition range," as the Los Alamos site was referred to at the outset. Colgate commented that the Army was putting "megabucks" into "what seemed . . . the worst place in the world to have a laboratory because there was no railroad and no water or any of those things that you normally need for a laboratory."[6] The last four seniors were quickly graduated, and the other students relocated. The federal government filed the petition of condemnation under the 1942 Second War Powers Act in the spring of 1943.[7] The price negotiations occurred that summer. In the end, the direct-purchase contract between the school and the Army was for $350,000, a middle ground between the school's asking price of $500,000 and the government's first offer of $275,000.[8] The school buildings and the campus grounds became Area A, the center of the Laboratory.

Fermor Church, the school's math and science teacher, tried to reopen an institution bearing the same name in Taos in 1944 but failed because recruiting students was difficult during the war. His school closed after a year, and he became an environmental activist and the founder of New Mexico Citizens for Clean Air and Water. He and his wife, Peggy Pond Church, had three sons who, ironically, all ended up working in the field of nuclear energy:

> While they didn't agree with her views of nuclear weapons, Peggy's sons did share the family's sense of loss at being evicted. Today [In 1995], the Churches still lament the loss of the school, still consider themselves victims of the war effort, still complain about family furnishings and collections never returned, and wonder what life might have been like had Oppenheimer never seen their land.[9]

The owners of Anchor Ranch, the second largest property on the plateau, also challenged the government's offer and settled for $25,000. The ranch became a test range for the gun program. Edwin M. McMillan recalled the state of it when he came to assess the place: "The owner of the ranch was moved out and he'd left everything behind. It was a complete ranch with house, barn, equipment, everything including a flat area, which would make a good test range, next to a small canyon so the control building could be down in the canyon with the gun on the flat above."[10]

This comment reveals the rapidity with which landowners had to vacate. Of all the evicted, those who had the least time to organize their departure were the smaller landowners. Daniel Lang, reporter for *The New Yorker*, commented on "the first settlers of this atomic colony" in 1948, writing, "These men were furtively invading Los Alamos to see what could be done with a mesa. The *only* obstacle to their taking over was A. J. Connell, operator of the Los Alamos Ranch School."[11] The Hispano homesteaders were completely obliterated from his account.

Aside from the Ranch School, the Army Corps of Engineers looked upon the land as unoccupied grazing land that they could acquire at little expense. When forty-year-old Marcos Gómez saw the first government surveyor on his family ranch, he said, "I think the government's coming." When his father Donaciano replied that these government officials could not come here he replied, "Yes, they can. The government can throw down our mountains here if they want to."[12] The homesteaders received very low compensations for their land and their grazing rights. Almost all of them signed quitclaim deeds. Those who objected to the amount offered found their claims deemed inconsequential by the Manhattan Engineering District. Three contesters are listed in the district's history: Elfego Gómez, Ernesto Montoya, and Adolpho Montoya, misspelled "Montoyo," a telling mistake, showing just how seriously their claims were taken by the district. While the school and other Anglo-owned property were acquired officially after long negotiations, the lands of Hispano inhabitants, some of whom had had family in the region for centuries, were taken much more easily owing to cultural and language barriers. Truly, "the small landowners held more than two-thirds of the privately owned land (and probably a similar proportion of the grazing rights); but they received less than eighth of the money."[13] While the school was paid $225 per acre, including the buildings, and Anchor Ranch received $43 per acre just for the land, the homesteaders received between $7 and $15 per acre with all improvements.[14]

More than fifty years later, the families and descendants of these early homesteaders entered a legal battle to press for justice and compensation. Executive Director of the Pajarito Plateau Homesteaders Association Joe Gutiérrez said the government's actions in 1942–1943 were "unbridled" and "due process was sacrificed for expediency and families were forced at gun point to leave their homes."[15] Furthermore, the owners of the school and Anchor Ranch were given time to evacuate—two months according to

Stimson's letter—but the homesteaders had to leave immediately, sometimes leaving everything behind. The case of the Gómez family is illustrative of the proceedings.

The Gómez family was at the ranch when two men from the Army Corps of Engineers drove up in a jeep. These men carried rifles and approached the family members, one of whom was in the field, planting. These government representatives told the family they would have to gather the possessions they could carry and leave by the next day.[16]

If a family's address was known, they should have received a notice, not of the condemnation proceedings but of the immediate taking of their land by the War Department. In theory, the homesteaders should then have been able to negotiate; but, instead, "an amount was deposited in court equal to the estimated value of the homesteaders' property, and title to that property immediately passed to the government." Marcos Gómez testified in 1998 saying, "What I remember is that they came and they decimated everything . . . buildings, *corrales* [animal pens], and homes. And from there they had us under guard. . . . They took us to where the school is."[17] With the land, the Gómez family, like the others, lost their livestock—several hundred animals—which, some homesteaders saw, were shot by the Military Police Corps. Donaciano Gómez's compensation check was sent to a fictitious address, and there is no record of him ever receiving it. Many homesteaders were listed under "address unknown."

There was a wide gap between the estimated value of these properties and the actual compensation. Enriquez Montoya's property of 62.5 acres, for example, was estimated at $17,500, but the Army offered compensation of $425. In the end, he obtained $1,250. However, because his address was listed as unknown, he was notified through a newspaper publication.[18] Today, a property of the same size in the town center where the evacuated land is located is worth about $250,000. The median real estate price is $230,276, "which is more expensive than 85 percent of the neighborhoods in New Mexico and 63.9 percent of the neighborhoods in the U.S."[19]

Contrary to the White Sands ranchers who had solid reasons to believe that they would be able to come back after the war, even if that promise was eventually broken, the evicted Los Alamos homesteaders had no knowledge of whether they would get their properties back. The ranchers kept a title to the land since they simply leased it to the government. The homesteaders, on the other hand, relinquished all their land rights. The main points in common between the two groups, however, are their inability to reproduce their self-sufficient lifestyles and the nostalgia for their past. Most families went to live in the Española Valley and went to work for the Laboratory in maintenance positions. After having lived off the land they became the blue-collar workers of the atomic bomb. Most saw the change as an improvement of their living standards, but others were nostalgic for their former life. Marcos Gómez and his wife, Maria, moved to Alcade where Marcos became a laborer and foreman for one dollar a day at Los Alamos, four miles away from his "El Rancho." In 1975, the couple was granted the right to visit the site of their old ranch accompanied by security guards. This area, Two Mile Mesa, had become a testing ground for detonators in high explosives research. The visitors were satisfied to see that the outdoor stone bread oven, pigpen, corrals, and chicken house also made of stone remained as markers of the homestead. Their reaction was highly emotional: "Both of us cried. We spent some of our best years there."[20]

During World War II throughout the United States, many public and private buildings and lands were put under control of the War Department and the Army to contribute to President Roosevelt's "arsenal of democracy."[21] General Groves later wrote in his autobiography that he was satisfied with the reaction of the local population to their project. He notes that the population's and state government's motivation in participating in the war effort was because a New Mexico National Guard regiment had been captured in the Philippines. He comments, "The support we received was superb."[22] Yet, although most plateau landowners patriotically accepted the sacrifice that the government asked of them, they nonetheless hoped to recover their land. Others refused to give it up willingly because they doubted the postwar outcome. They waited for the Army, gun in hand, but their stories did not make the headlines, contrary to those of ranchers John Prather and David McDonald: "Ted Mather's wife sat on the front porch of their homestead with a shotgun, indicating how they felt about the taking. Some Hispanic landowners in Los Alamos considered refusing to sign a quitclaim deed because they believed that the government was going to take the land whether or not they signed. They hoped that not signing would give them better legal grounds for reclaiming land after the war."[23]

Some of these displaced landowners saw their family land being transformed into a testing ground for explosives or into the town's golf course, park, and cemetery. When the land was condemned in urgency, no reports and no studies were undertaken to evaluate what might be the impact of a weapons' laboratory and development of an urban environment on these areas, which included the game refuge, sacred Pueblo burial grounds, and the nearby Bandelier National Monument, which sheltered invaluable archaeological treasures and later served as a recreational park for the Project's scientists.

Everything is "Picturesque" on the Frontier of Science

The first group of scientists recruited by Oppenheimer arrived at Los Alamos in March 1943, and on April 1, Los Alamos had officially become a military establishment under the supervision of Oppenheimer on the civilian side, Colonel John M. Harman on the military side, and the University of California as a War Department contractor.[24] The quickly built community of a few hundred grew to more than six thousand by the end of the war and thrived because it combined two important factors for success in a difficult isolated environment: technology and government capital.[25]

At the beginning of Project Y, several scientists and observers such as Sterling Colgate, the LARS student who had recognized the visitors, commented on the idea that the western setting was the least expected place for such a "futuristic" enterprise. The prevalence of western mythology in the American mindset had long prevented any association of ideas of modernity with the West. The "idea of a modern West" seemed to be an "oxymoron" to many people because "the 'real' West can't be modern."[26] In New Mexico, the contrast is amplified because the state is a concentrate of western mythology that derives from its landscapes, its native populations, and its historical landmarks. The state was the stage of Kit Carson's campaign against the Navajo nation, of the arrest of Billy the Kid by Lincoln County Sheriff Pat Garrett (1880–1882), and of the frontiersmen's journeys along one of the most famous wagon trails westward, the Santa Fe Trail. Just the name Santa Fe summons pictures involving tumbleweeds, colorful Indian rugs, and cowboy boots.[27] New Mexico is the home of frontier legends, and the tourist industry has always sought to capitalize on this heritage to guarantee visitors a unique western experience. The New Mexico Tourism Department, for instance, boasts of the state's four hundred ghost towns of the late 1800s.[28]

Such a concentration of signifiers accentuated the propensity of participants in the atomic project to view New Mexico and their adventures there through the lenses of western mythology and Manifest Destiny. General Groves, for instance, was fascinated with the western mystique because he had spent most of his life on Army posts, dating back to the American Indian Wars with his father who had been an Army chaplain. He had met men who had taken part in the final chapters of the conquest of the West in the 1890s. He thus saw in the dawn of the atomic age the opportunity to expand on the "win the West" incentive and find a new frontier in science.[29] Groves reportedly told Colonel J. C. Marchall that the Los Alamos scientists "will like anything you build for them. Put up some barracks. They will think they are pioneers out there in the Far West."[30] Imagination and mythmaking can help create cohesion among people of various backgrounds, who did not partake in a nation's great experiences but adhere to its myths. Richard Slotkin defines a myth as the description of "a process credible to its audience, by which knowledge is transformed into power; it provides a scenario or prescription for action."[31] This definition means that myths have the ability to dictate the behavior and relations of people who are influenced by what they imagine a certain environment to be. The attitudes of the scientists and their families during the wartime years of Los Alamos are a fascinating illustration of this process.

Western mythology thus provides a second "familiar channel" (after the military) through which changes were brought. When the first scientists arrived in the Jémez Mountains, their knowledge of the region was most likely based on popular, prototypical imagery of the "Old and Wild West." In their accounts, many Hill dwellers, as Los Alamos' residents were typically called, made the analogy between their experiences and tales of Frontier life. Ruth Marshak, for one, writes, "I felt akin to the pioneer women accompanying their husbands across the uncharted plains westward, alert to danger, resigned to the fact that they journeyed, for weal or for woe, into the Unknown."[32] Even the scientists' Sunday pastimes reflect their taste for the mythic past. One of their hobbies was mining, which is still popular among tourists today. They would study mining reports and geological maps before hiking to old mines hoping to find gold or any other precious materials. Jean Bacher, the wife of physicist Robert Bacher from the Massachusetts Institute of Technology (MIT) Radiation Laboratory, writes, "From the passion of exploration it was only a step to the mining mania. Mining offered not only exercise but also the spirit of romance of the Old West."[33]

As they had not chosen to go to northern New Mexico for the appeal of the location but to fulfill a mission, most scientists and their families knew very little about the area. While looking at maps to locate Los Alamos in the Jémez Mountains, John Manley could not find it because, following Oppenheimer's oral instructions, he was looking for a place he thought was called "Hamos." His phonetic spelling of the Spanish name *Jémez* shows how foreign the place must have seemed to him.[34] The shock was even more violent for incoming Army personnel who were sometimes so bewildered that they "figured they had somehow landed in North Africa where the War was raging at the time."[35]

Other residents showed their ignorance of the local environment through their attempts to transfer features from home onto the mesa, including luxuriant gardens that were near impossible to maintain in the arid climate. It was only thanks to the immense technological means the site officials had at their disposal that they managed to maintain this water-greedy community. Kathleen Mark, the wife of physicist and mathematician J. Carson Mark, recalls, "Whether out of nostalgia or just a dogged will to succeed, some people tried year after year to create gardens. . . . Then by dint of much watering they sometimes achieved rows of green seedlings and occasionally knew the triumph of actually eating the fruit or, in this case, the vegetables of their labor."[36] Mark's anecdote reflects the frustration of her fellow expatriates who sought to bend the harsh environment to their will, like frontiersmen before them.

The West as an imaginary place on the frontier of civilization also conditioned the way scientists interacted with the non-English-speaking locals who were hired at Los Alamos; the newcomers were predisposed to consider the locals with inquisitiveness and paternalism. The Project hired profusely from nearby communities to remedy the labor shortage and to comply with secrecy and security rules: construction workers, janitors, custodians, cooks, clerks, and housekeepers from Española, Chimayó, Bernalillo, and Las Vegas, New Mexico, as well as from the nearby Indian pueblos of San Ildefonso, Cochití, Tesuque, and Santa Clara. The overcrowding of the Hill made it necessary to hire people who would not have to live there. This brought about novel contacts between two antipodal cultures: Spanish-speaking and Indian rural workers on the one hand, and highly educated American and European scientists from the world's greatest universities on the other. The encounter aroused curiosity on both sides, which was favorable to exchanges, but the difference in status also fueled stereotyping. Thus, Cleto Tafoya, a former governor of Santa Clara Pueblo who was hired as a cook, could be seen pouring soup at the East Cafeteria. Residents called him "chief" because they could not pronounce or remember his name.[37] No matter their status in their communities, local workers had an inferior status to Project scientists on the Hill, and their inferiority combined with certain preconceptions made them prone to caricature.

Charlie Masters, a teacher at the Los Alamos school, recalls a skit staged at a British Mission party in which Otto Frisch played the part of an Indian maid wrapped in a rug. His portrayal of an indolent, slow-working employee who cleaned dishes on window curtains and drank alcohol as a reward is one example of the simplistic views some Los Alamos dwellers fostered of their Indian maids.[38] The following caption can be found under a picture of Frisch on stage: "This exaggeration took to the point of hilarity the comparison between regimented routines of the New England, Southern, or Midwestern transplants and the ageless, unhurried rhythms of the Indian natives."[39] The caricature underscores the discrepancy between the representatives of progress and those of immobility. Indigenous cultures were at the time represented in America's history as a reference point, a control sample to evaluate and enhance the speed and extent of the nation's progress. It can be pointed out that disregard for cultural alterity reaches its paroxysm in the display of the jumbled use of signifiers: the "maid" is wearing a *concha* belt around his neck. *Conchas*, meaning "shell" in Spanish, are oval pieces of jewelry usually in silver and turquoise attached to braided leather to be worn around the waist.

Eleanor Jette, the wife of Eric Jette, who came to work at the metallurgy lab to study plutonium, recalls how shocked she was at the poverty in the Pueblo villages and remarks on the economic divide between the two groups. To families who came from a society of abundance, Native communities seemed to be barely surviving. She writes, "There was abundant evidence of the poverty that oppressed the pueblos for centuries. I realized that even the meager wages the maids received were riches to most of them."[40] Maids were paid $3 a day, $1.5 for each half day, and used the money to improve their homes or to buy food at the Los Alamos commissary.[41]

The most salient comments about the Indian population concern their value in terms of entertainment: they were "picturesque." Local workers brought a kind of exoticism that brightened the expatriates' experience. Ruth Marshak expresses this idea when she writes, "They [the native New Mexicans] gave a remarkable flair to the place." She also notes the incongruity of "the oldest peoples of America, conservative, unchanged, barely touched by our industrial civilization"—a definition that recalls the Manifest Destiny trope—participating in the birth of the atomic age: "The Indians and Spanish-Americans of New Mexico were the most unlikely of all peoples to be ushers to the atomic epoch."[42] Her comment, like that of LARS student Sterling Colgate, reflects how unnatural the association of the West or its inhabitants with modernity appeared to easterners.

The word "picturesque" occurs time and again in the narratives of early Los Alamosans such as Elsie McMillan's, which described driving on the road to the site. They drove "past Black Mesa, which seemed to me like a wonderful sentinel, guarding the Indian pueblos so near. The chamiso and the tumbleweeds going along made this wonderful country even more *picturesque*."[43] Other Hill dwellers similarly rejoiced in the attractive, photogenic location of the laboratory, using the same adjective to express the charming setting that resembled a painting or a western film. Emilio Segré, for instance, chose a small isolated log cabin as his personal, secondary laboratory to measure spontaneous fission. The cabin had belonged to a rancher in a valley a few miles away and was in "one of the most *picturesque* settings one could dream of."[44] Again, one notes the powerful tropism of western iconography.

Phyllis Fisher describes her first encounter with her house help, Apolonia. She was "a short, middle-aged, stooped Indian woman from a nearby pueblo" who arrived "rolled up in a bright red blanket and all smiles" and called her "Meesie Feeshah." Her thought on the colorful appearance was that, whether she was a competent worker or not, she would "be worth her wages in entertainment value alone. She is sweet and *picturesque*, and I do love to watch her. If she does nothing more than stand around, I'll find my housework less boring."[45] Her fondness for her maid, therefore, did not proceed from Apolonia's character as an individual but rather from what she projected as the embodiment of an exotic world that her employers felt privileged to witness. In a letter to her parents, Fisher also writes of her surprise at discovering that a Pueblo was "clean" and "attractive."[46]

The men were not depicted in such detail as they were more taciturn and wore work clothes that were less colorful. Their unusualness came from the way they wore their hair. Berenice Brode uses the same term, "very *picturesque*,"[47] to describe the Indian workers commuting to the Hill and emphasizes the entertainment value of Indian dances: "Whatever else they meant to the Indians, [they] provided good fun and a show for everyone. Our people at Los Alamos provided a good and enthusiastic audience and the Indians liked it."[48] The cultural meaning of signifiers is again disregarded in favor of the amusement provided by the scene, hence, perhaps the need to add an exculpating "they liked it." The constant reference to picture-like charm is new evidence of how ubiquitous representations of the Old West were in these pioneers' minds, most of which probably came from the myriad of epic Western movies released between 1920 and 1940. The symbol of the noble Indian warrior and the belief in Indians' inexhaustible adaptability to change produced descriptions such as that of Charlie Masters:

> They [the Indians] had a ponderous, undisturbed quality which
> made us remember that their kind had endured through tribal
> wars, drought and famine, Spaniards and slavery, Yankees and
> machines, and that assuredly they would have no trouble surviving
> the atom-smashers. Unhurriedly and with the minimum of adapta-
> tion, they altered the manner of their living temporarily to serve the
> Hill as maids and waitresses, as janitors, firemen, and cooks. But no
> matter what the nature of their work, their native dignity remained
> unimpaired.[49]

Likewise, Berenice Brode mentions their "serene dignity," which made them seem "more like guests than servants."[50] From the other side, though, even if local workers rejoiced in their new economic welfare, they were aware of this status as "servants." It set the basis for their relations with the Hill.

The fascination Hill dwellers developed for Puebloans made them eager to possess tokens of their cultural discovery. Whether it was by exchange of gifts, a simple touristic transaction, or sometimes through theft and violation of local archeological sites, many Hill dwellers acquired souvenirs to display on the mantles above their fireplaces. An unprecedented demand for items such as rugs and pottery put a strain on the potters of Santa Clara and San Ildefonso. Charlie Masters calls this appetite for arts and crafts the "going-native process." This form of artistic primitivism was an antidote to the stress of life on the Hill. For the same purpose and to facilitate encounters, people organized folk-dancing groups with the Old Timers Square Dance Club. The people from Los Alamos would go to San Ildefonso to watch Indian dances, and, in turn, they would invite Indians to participate in their dances. At the Pueblo, "It was the Yankee invasion all over again," Masters wrote. They brought wieners, buns, Coca Cola, record players, electric cords, generators, and guitar and violin players in case a generator did not work. They shouted instructions to each other in front of their Indian hosts who observed them "silently and sympathetically."[51]

Berenice Brode argues that their influence on the surrounding communities was a positive one. She opposed Santa Feans who accused them of disrupting the Natives' traditional lifeways because she had learned from her Indian acquaintances that they liked the people on the Hill on the grounds that the relationship was chiefly work based. She comments, "It was the first time they had known any group of Anglos who were not primarily interested in their welfare or curious about their cultural patterns."[52] She believed the relations to be the same for those who wove ties with Spanish families. Yet, the presence of Los Alamos visitors did induce some modifications to the rhythms of pueblo life. The Indians felt the urge to organize more dances and added festivities on Sundays to meet the demand of scientist families who came to relax on weekends.[53] Hill dwellers suspected that the Indians were inventing new, nontraditional dances for them, knowing that they would not notice the difference. Brode argues that the Pueblo Indians suffered from pressure to remain rigorously traditional for the benefit of tourists but by asking for dances and crafty souvenirs, visitors from the Hill automatically participated in the intensification of such pressures.[54]

Brode counters the attacks of local anthropologists by asserting that the influence was reciprocal. While people in Los Alamos acquired Indian artifacts to decorate their homes, Indians used their wages to buy comfort additions to theirs (e.g., furniture, a bathroom, a new floor). She explains, "We just enjoyed the Indians and what they had to offer, just as they enjoyed our Hill life and drinking Cokes in our PX." Exchanges were, therefore, straightforward supply-and-demand transactions. When she visited the pueblos again in 1948, she noticed the modernization that had taken place. She thought it did not have much to do with the presence of the lab but more with the fact that the local residents all earned wages. The latter was, of course, the result of the former. By that time, surrounding villages were economically connected to the Hill. The Hill's influence over Santa Fe is presented in the same lines. The new residents got local businesses going and the impetus, despite shortages sometimes caused by their overwhelming presence, helped the town prosper. They had "livened up" this "ancient sleepy town." Brode doubts their "presence affected the *picturesque* atmosphere well established in Santa Fe."[55]

Phyllis Fisher's perspective slightly differs from Berenice Brode's. Fisher describes her gradual adaptation and infatuation with the character of the region she calls an "almost foreign land." Her husband, Leon Fisher, was an instructor at the University of New Mexico in the Department of Physics and Astronomy for a few months before being recruited by the Manhattan Project. After moving to the Hill, she became interested in the art, the philosophy, and traditions of local people. She later pointed out in retrospect that "while enjoying their entry into our world," the transplants never thought of how their entrance was perceived. "It never occurred to us that we were intruders into their world, or that we behaved toward them in a condescending manner," she writes. She continues, "In fact, it was a while before we knew these gracious people as individuals and as friends and not as curiosities." To exemplify this self-aware thought, she recalls the day they hiked on Black Mesa, a sacred place to the Pueblo peoples. The hikers became aware of the intrusion once they had reached the summit and saw evidence of a shrine. Although she came to question her behavior toward Native Americans, Fisher concurred with Bernice Brode in saying that the cultural influence between the two groups was mutual. While the Anglo population purchased pottery, learned how to make fry bread, displayed Navajo blankets in their homes, and wore Indian jewelry, the native population bought sets of dishes from their Montgomery Ward catalogues, made peanut butter sandwiches, put linoleum on their floors, and wore jeans.[56]

Perhaps what these analyses failed to perceive, though, is that when two cultures are in contact, one being the dominant culture in society and the other being on the margin, the former gains from the encounter by the observance of uncommon lifestyles, whereas the latter is confronted with the risk of being engulfed through acculturation and economic penetration. So, despite evident reciprocity, these cultural exchanges did not mean both cultures were to be equally impacted.

In 1995 Katrina Mason interviewed individuals who had grown up in Los Alamos for her book *Children of Los Alamos*. Some recalled with affection their Indian caregivers and housekeepers who gave them Tewa names.[57] They remembered the school fieldtrips to see the potter Maria Montoya Martínez. One of the interviewees, Martha Bacher Eaton, linked her appreciation of spirituality to her relations with the Indians. She said, "I don't think that I would have been this kind of a spiritual person if I hadn't run into the Indians. . . . There wasn't any way to talk to them, so I had to relate to them on a different kind of dimension."[58] She was inspired by the Puebloans' silence with outsiders and with each other. One notes in Bacher Eaton's and previous testimonies what seems to be a tendency of Anglos to interpret Indian attitudes positively, with admiration or even idealization. Considering them as characters in an idyllic picture was not unrelated to this tendency. One counterexample was Edith Warner, who sometimes attended the parties where scientists and Native Americans socialized, with her Indian companion Tilano. Peggy Pond Church observes that Warner "resisted the temptation that many white people feel to idealize the Indians, the effort to find in this ancient culture all that seems lacking in our own."[59]

Some of the Los Alamos children, like Edith Warner, fell deeply in love with New Mexico and came back to settle there. Jim and David Bradbury, the sons of Norris Bradbury, who was the second director of LANL, both came back to work there—the former as a physicist and the latter for the New Mexico State Land Office in Santa Fe and then at the environmental restoration at LANL. Jim was friends with a local Hispano, Secundino Sandoval, and explained that he envied the sense of belonging and the ties to the land that he perceived in the extended Hispano families. Once he came back, he found that he had been missing these things. He says, "The sense of connection to the land. The space, the confluence of these cultures . . . The things that stand out in my mind are these relationships with the Indians. They were much more than maids. They were surrogate parents in our case."[60] He reconnected with Isabel Atencio, the Bradburys' housekeeper.

His brother David also remained close to Isabel and visited her yearly at San Ildefonso.[61]

Through the Eyes of Others

Thanks to oral histories, memoirs, autobiographies, photographs, and the work of historians, substantial knowledge has been acquired of the early scientific pioneers' perspectives on the world and the people surrounding them at Los Alamos. Accounts by scientists and their wives allow us to have a better understanding of what their feelings were toward Pueblo Indians and Hispano residents of the area. On the other hand, only rarely do published works provide an indication of what these locals' feelings were toward the newcomers. Katrina Mason gathered the recollections of two New Mexican children in her book: Severo Gonzales, whose family had homesteaded on the Pajarito Plateau, and Dimas Chavez from Torreón, a village behind the Sandia Mountains.

After working for LARS, Severo's father, Bences Gonzales, turned to work at the commissary in Los Alamos; this job enabled his family to stay on the mesa even after they had to give up their land to the Army. Severo Gonzales graduated from Los Alamos High School in 1948. He shared his impression of the people who came to work on the Manhattan Project with humor, stressing the ludicrous effects of western mythology on the newcomers:

> [They thought] they were going where everything was wild, [where] you might be shot with a bow and arrow by an Indian. . . . We were poor but we always had everything we wanted. A lot of the students [when] they were graduating, they'd give us their tennis rackets, balls, skates, hockey clubs, pucks . . . Boys that were going back East, they had money—real nice jackets these boys would give us. . . . If Los Alamos wasn't what Los Alamos is, this whole valley would be about a fourth of what it is. Santa Fe would be half of what it is. It would be selfish to say I wish I was still homesteading.[62]

The pragmatism, bordering on fatalism, in Gonzales's last sentence shows how straightforward and devoid of embellishment the locals' perception of the lab's presence was, a glaring contrast to the romanticized frontier tales of Los Alamosans. New Mexicans quickly made a clear distinction between the economic windfall from the Hill and what they were culturally attached to in the Valley.

Dimas Chavez's family moved to Los Alamos in August 1943. Chavez recounted how his father, a farmer and rancher in Torreón with an eighth-grade education, had had to look for a job as a laborer in Santa Fe after most of the crops had withered in the early 1940s drought. In Los Alamos, he first worked as a heavy-duty operator and then was involved in water and sewage treatment. Dimas, who was one of five children, did not speak English at the time; he remembers, "For some reason in the early years, there were few Spanish-speaking students or residents. There were numerous Spanish-speaking personnel who worked there, but they all commuted from down in the Rio Grande valley."[63] His mother made a deal with the scientists' wives: she would teach them how to cook New Mexican dishes if they taught her son how to speak English. After graduating in 1955 and leaving for Colorado to become a telegrapher, Chavez returned to Los Alamos, took a job at Metzger's gas station, and then became a truck driver for the Lab. He explained being traumatized by the hardships of going to school with smart children because he had to work twice as hard to keep up. Because he became afraid of this academic atmosphere, he did not go to college until 1956. Finally, in 1972, he went back to work at the Lab.[64]

Most of the interviewees of the University of New Mexico Oral History Project, "Impact Los Alamos" in the Española Valley, also started working at Los Alamos after the war—after the Zia Company became the Lab's principal subcontractor in 1946. The few people who remembered the wartime years, whether they were workers hired by the Laboratory or children whose parents worked for the Project, talked about it in a comparative perspective. There was a "before Los Alamos" and an "after Los Alamos." Contrary to the expatriates who recalled the novelty of the people and environment around them, the locals' memories focus on how drastically their life was changed. One of the oft-mentioned changes was that New Mexicans no longer had to leave to find work in other states but could stay at home with their families. Children could see their parents every day. Bernadette V. Córdova, for instance, recalls her father taking "the G.I. bus to the Hill" during the war and the chocolate bar that she would get once a week on Fridays.[65] José Benito Montoya remembers his father coming back home on Sundays from the Hill where he worked as a carpenter. José Benito was accustomed to him being away because he had worked in Colorado before the Manhattan Project; but they began leading a better life after the arrival of the Lab. Instead of making their own pants and shirts, they could buy clothing at the store.[66]

Old-timers remember how the Army trucks would travel a fifty-mile route through Chimayó, Española, San Ildefonso, Santa Clara, and Tesuque to pick up blue-collar workers on the plazas of Pueblo and Hispano villages in the early morning and take them back after a ten-hour workday. Not having to pay for rent or transportation was another advantage of living in the vicinity. Joe G. Montoya lived in the barracks in the summer of 1944; but in 1945–1946, he started commuting with his friends from the Valley, drinking beer on the way down.[67] Some focused on the wages they brought home as they were considerably higher than the standards of the region. The increase made a great difference according to Richard Cook, the head of the Bond mercantile company, the largest employer in Española at the time. He supplied Los Alamos contractors with lumber, plywood, sand, and gravel for their construction work. It was an unprecedented market for this kind of business in the area. Many observers in the Valley concurred to say the economic spin-off for local businesses was impressive. Los Alamos was a financial goldmine.

Cook saw the effects of the Los Alamos payroll in the Valley. He notes people could buy better houses and ensure a better education for their children, who then would not have to leave to find employment. Like Severo Gonzales, Cook says, "If it weren't for Los Alamos and the payrolls providing the economic base that is what keeps everybody glued together, I don't know what we would have done . . . We would probably be a sleepy little town like we were number of years ago." The use of the term "sleepy little town" is interesting as it shows that New Mexicans had also integrated some of the prototypical concepts and signifiers of the mythical West. When talking about the people living on the Hill, Cook admits the newcomers seemed to feel superior because "they were very proud to be Ph.D.s and would look down on people of the Valley, considering them a little bit inferior but they were a very small minority." He adds, "Even if they tried to impress Valley people, in most cases, their practical knowledge wasn't as good as their book learning, so . . . most people ignored it."[68] Delfido Fernández explains that if there was always some discrimination, it was because Hispanos were the minority on the Hill; he comments, "It was natural Anglos were leaders: that was the way it always was. Anglos thought they were superior."[69] One observes the same sense of fatalism and the internalization of the invader's point of view. The conquerors had a "natural" right to lead the way toward modernity.

The first generation of workers had experienced the hardships of agrarian society, separations while seeking work elsewhere, and the general poverty in the villages of northern New Mexico. And then they experienced the arrival of the Lab, which brought them back to their homes, gave them a reprieve from backbreaking days of subsistence farming on worn-out land, and helped them purchase new comforts or cars. They saw Los Alamos as the savior of their Valley and a stabilizing change in their lives. With their grateful sentiments, New Mexicans initially accepted their role in the scientific conquest. The memory was passed on to the next generations. Even those who were too young to recall their parents leaving the state knew that, were it not for Los Alamos, they would have had to leave, too. Senni A. Gallegos talks of this heritage. As child, she did not have a friend whose parents did not work on the Hill, mostly in the janitorial service or cleaning people's houses. She asserts,

> If Los Alamos hadn't been there, who knows what would have happened to us? We would probably have to move. Everybody working there were very fortunate that they were able to remain in the state. We could have made gardens and be self-sufficient by selling our produce, but I think that people thought that was too difficult, too hard, like slaving yourself to death. One of my neighbors couldn't get a job up here, so he had to go away in California. Before a lot of people did that: leaving their wives behind. And when Los Alamos came here, that's when people stabilized.[70]

On the other hand, interviewee Genaro Martínez recalls with bitterness his days as a blue-collar worker at Los Alamos. In a stratified community where social classes were based on educational attainment, the fact he had no diplomas soon became an issue. Martínez started working there in 1942 at age fifteen, as a laborer and truck driver transporting coal and wood, for about fifty cents an hour. His desire was to be hired as a carpenter, since he had acquired the skills while building houses with his father, but he could not secure this job. He remembers, "I didn't get a good job or a good salary because I didn't have the education. I saw people from all over the U.S., they were not educated but they were Anglos and they would get the jobs. . . . I worked as a janitor for 25 years. I couldn't move up without that high school diploma. We were the lowest pay."[71] He organized a union in 1958 to defend pay rights and denounce the treatment from some of the supervisors.

Unfortunately, there are no records of the Pueblo Indian maids' and workers' experiences in the Lab's early years. While Hispanos are well represented in the series of interviews upon which this book is based, Pueblo Indian workers are conspicuously absent. In his introductory speech at the "Impact Los Alamos Symposia for the Community" in Albuquerque, Carlos Vásquez, the principal investigator of the "Impact Los Alamos Project," explained that the two weak points in their research were the representation of women and of Native Americans.[72] He accounts for the absence of Native American interviewees in an ensuing article, saying that they "were extremely reticent to be interviewed." In the mid-1990s, the Department of Energy and other government agencies conducted interview projects among the Pueblo Indians. "Perhaps, the Pueblos identified our work with those efforts," Vásquez thought.[73] The Project team observed that people were nervous to talk about Los Alamos.

During my own visits in New Mexico, I made the same observation that Los Alamos was a highly sensitive topic. Darryl Martínez, a resident of San Ildefonso Pueblo, talked to me about his grandmother Philipita Torres, who worked as a maid in Los Alamos in the 1940s. She had told him that the Army trucks would come pick her up on the plaza next to the large tree and that she enjoyed working there until the end of her professional life. The people whose house she cleaned treated her nicely. She would invite them to watch the pueblo dances. When Darryl Martínez talked about Los Alamos in the 2010s, he insisted on the fact that it was no longer the same because fewer people could work there. Getting a job had become more difficult. According to him, the members of later generations no longer worked for the families or the contractors but applied for positions at LANL directly; however, these jobs were scarce and required qualifications that were more difficult to obtain. His uncle, for instance, worked in the Tribal Relations Department at the mail office, where he oversaw the relationship between the Lab and surrounding pueblo nations.

According to Martínez, most people in his community would agree that the Lab had benefited them in terms of employment but that the situation had deteriorated.[74]

Local communities were indeed the privileged, almost exclusive, beneficiaries of an abundant employment source when the Project started; but once the secret town opened its gates to the rest of the world, abundance was replaced by increasing competition. Darryl Martínez's analysis of this evolution reflects the same wistfulness for the blessed time of job profusion that was expressed by the Hispano interviewees of the Española Valley. Yet, the nature of the positions that were unofficially reserved for locals during the wartime years should be underscored. They were a category of maintenance, low-skill, and sometimes menial jobs to which the poverty-stricken population, eager to find a way to remain with their families, was grateful to have access.

Many interviewees, such as Richard Cook and Genaro Martínez (and others who will be mentioned later) expressed a sense of inferiority because they saw the types of jobs locals could do in light of those reserved for highly educated outsiders. From the early days of the Manhattan Project, the principal criteria separating social classes in the peculiar scientific community were education and the number of degrees—or even Nobel Prizes—one held. At school, children had different opportunities depending on their scientist fathers' clearance levels. Interviewee Nella Fermi Weiner recalls,

> [Los Alamos] was a very stratified society. . . . There was a real educational gap. . . . There were the kids of the physicists [and the other kids]. We were not necessarily more intelligent but we certainly had more opportunities than these other kids. We put it down on our intelligence, but looking back I'm inclined to say we really had better chances . . . not only have gone to better schools but also having parents who were, to say the least, literate.[75]

As new generations of New Mexicans, more educated than their parents, aspired to better positions at the Laboratories, they faced the difficulty of challenging the established order. Meanwhile, the fascination of inhabitants of the Hill for the idyllic cultures surrounding them did not subside. Words were even borrowed from the pueblo culture to be incorporated into the scientific culture: *Kivas* came to designate the laboratories in

Pajarito Canyon where scientists conducted criticality research and streets were named San Ildefonso Lane, Tewa Loop, and Santa Clara, Navajo, or Ohkay Owingeh Places. Taking the cultural uncanny to its extreme, during Operation Redwing in the summer of 1956, explosions were named after Native American peoples, including four nations living in New Mexico— the Zuni, Apache, Navajo, and Tewa. The Tewa device detonated on a barge, producing radioactive fallout that contaminated Bikini, Enewetok, and more than two thousand square miles of ocean.[76]

Five

The Atomic Sun Shines Over the Desert

The acquisition of vast portions of land in New Mexico at the onset of World War II was a major asset to the Manhattan Engineering District. First, it could rely on readily available military help for security, protection, or transportation. Second, the militarized landscape proved profitable to field-test the plutonium device before considering its use in warfare. In March 1944, the Ordnance Engineering Division put Harvard physicist Kenneth T. Bainbridge in charge of a team to make the preparations for the test. While looking for an adequate location, General Leslie Groves imposed just one criterion: there should be no Indian presence. This was in accordance with his previous misgivings about problems he might have with Secretary of the Interior Harold L. Ickes, who had jurisdiction over the Bureau of Indian Affairs, and John Collier, commissioner of Indian Affairs appointed by President Franklin D. Roosevelt in 1933.[1] A supporter of the "Indian New Deal," Collier had endorsed the Indian Reorganization Act of 1934 that allowed tribal self-government and consolidated individual land allotments back into tribal hands—a reversal of the 1887 Dawes Act. Collier had first encountered Native Americans in Taos in 1920 and had written in 1938, "So intimately is all of Indian life tied up with the land and its utilization that to think of Indians is to think of land. The two are inseparable."[2] Throughout the testing operations, Groves was particularly attentive to possible legal outcomes, such as being sued for damages on civilian structures, and the dealings with the local ranching population followed his line of thought.[3]

Bainbridge, now Trinity Test director, was in charge of choosing a site; he looked for a place with a flat terrain and good rail transportation, at a reasonable distance from Los Alamos that would enable workers and supplies to travel relatively easily, as distant from any human presence as possible to make sure people could be evacuated rapidly, and to reduce the number of witnesses to a minimum. David Hawkins, a friend of Oppenheimer who became the official Manhattan Project historian, believed he was the first to suggest the region of La Jornada del Muerto for the test. Born in El Paso, Texas, but having grown up at La Luz near Alamogordo, he recalled in an interview in 1995 how, as a child, he had wandered "all over the Tularosa Basin, one way or another, looking for minerals, looking for excitement, looking for rattlesnakes, looking for adventure of the desert kind."[4] The story echoes that of Oppenheimer's suggestion of the Pajarito Plateau to build the atomic laboratory.

Despite ranking second on the list of potential sites, which included two others in the state (the lava beds, now the El Malpais National Monument south of Grants, and an area between Cuba and Thoreau[5]), the area had the advantage of closeness with Los Alamos (230 miles), suitable weather, seclusion, and the fact that it already belonged to the military. All General Groves had to do was to ask the Air Force, which controlled the Alamogordo Bombing Range then connected to the Alamogordo Airfield, for permission to use an 18-by-24-square-mile portion of the desert. This overlooked the issue of local ranchers. The test necessitated forbidding all grazing permits around the site, and some herders left cattle on the range out of defiance. Gerard J. DeGroot called these animals "the first martyrs of the atomic age."[6] Some of the ranchers' houses earned national fame in their role during the Trinity operation. The bomb was assembled at George McDonald's ranch house two miles south of Ground Zero, and the soldiers were housed at his brother's, Dave McDonald, who was the father of interviewee David McDonald. The first McDonald Ranch, now part of the memory of Trinity, was transformed into a museum where visitors make a first stop on their way to Ground Zero.

The town of Albuquerque also participated in the project by receiving vast quantities of material that would then be used to conduct and monitor the experiment. These shipments were addressed to a Mr. J. E. Burke (the name could be a reference to the short form of Albuquerque: "Burke"), Department of Physics, the University of New Mexico. The supplies were then freighted by train to a place on the outskirts where Army engineers collected and trucked them down to the site.[7] A local business, New Mexico's largest construction corporation run by Ted Brown in Albuquerque, was chosen to oversee the construction of the site; two hundred workers were brought to build roads and temporary dwellings.[8] A scientific frontier town mushroomed in a few months, and by mid-July 1945, the site housed 250 scientists and technicians, and an equal number of soldiers.[9]

Witnesses to the Dawn of the Atomic Age

As the test neared, stress reached such a level that Groves requested psychiatrists "to assist scientists in keeping a mental equilibrium."[10] Enrico Fermi famously attempted to release the pressure by organizing a betting pool on whether the bomb would ignite the atmosphere, destroy New Mexico, or devastate the whole planet. The mighty detonation on July 16 exceeded all expectations. After the blast, the scientists' reactions varyingly attempted to express the significance of the awe-inspiring event in countless metaphors and superlatives. The two most famous are Oppenheimer's quote from the *Bhagavad Gita*, "I am become death, the shatterer of worlds,"[11] and Kenneth Bainbridge's declaration, "We are all sons of bitches." General Thomas Farrell, deputy to General Groves, summarized the effects of the atomic spectacle in a list of adjectives, "unprecedented, magnificent, beautiful, stupendous and terrifying."[12] They were only the most famous witnesses to the blast. Max Coan, a soldier who was present, has associated that moment with his attachment to the region:

> The magnitude of the occasion etched more deeply the vivid picture of New Mexico in my memory. . . . the atomic bomb debut is just another one of the things I enjoy recalling about the Land of Enchantment. . . . I can't remember when I first started liking New Mexico, but it must have been that first day when I stepped off the troop train to the hot platform of the railroad station in the quiet little city of Alamogordo.[13]

Apart from soldiers and scientists, unexpected individuals witnessed the explosion. Up to 250 miles away from Ground Zero, people heard, saw, or felt the explosion. In Gallup, 235 miles away, houses shook and windows blew out. Years later, New Mexican newspapers published the testimonies they had been banned from publishing on that day. The Manhattan Engineer District (MED) had planned the release of articles in the press that would justify the event to the public without risking any security breach on the true nature of the explosion. Defense Department officials had prepared six different stories, including a scenario in case of civilian deaths. Newspaper publishers of several major local publications had received Federal Bureau of Investigation (FBI) agents on the next day who suppressed all stories that neared the truth.[14] Meanwhile, the radio in Albuquerque broadcast the news that an ammunition storage site had exploded in the area of Alamogordo.

The *Roswell Daily Record* published only a small article under the headline "Ammunition Blasts at Magazine on Alamogordo Field." The newspaper believed the light was caused by the descent of a meteor and had been preparing to publish the much more sensational "See Great Blaze of Light in Sky," but they were censored. The news was kept out of the media in the East, and most of southwestern residents believed the official version. Nevertheless, New Mexican witnesses later claimed that they were not fooled by the cover story that did not make any sense to them. Fritz Thompson from the *Albuquerque Journal* reported, "Several ranchers say they never believed the Army cover story. . . . But they didn't guess what it was until the devastation of bombs at Hiroshima and Nagasaki weeks later. Even then, they didn't guess the importance of what had been wrought in their backyard."[15] Newspapers and wire services were swamped with questions from local observers anxious to know what had happened or to send a warning. Near Silver City, forest rangers called a nearby observatory to see if there had been an earthquake. Some people thought they had heard multiple explosions.[16] The Bulletin in the *Albuquerque Tribune* on July 16, 1945, read:

> Several inquiries have been received concerning a heavy explosion which occurred on the Alamogordo Air Base reservation this morning. A remotely located ammunition magazine containing a considerable amount of high explosive and pyrotechnics exploded. There was no loss of life or injury to anyone, and the property damage outside of the explosives magazine itself was negligible. Weather conditions affecting the content of gas shells exploded by the blast may make it

desirable for the Army to evacuate temporarily a few civilians from their homes.[17]

Thus, despite secrecy measures, the first atomic blast did not go unnoticed in New Mexico. It did not go unnoticed in western Texas either, where there were a few sighting stories too. The story that became most famous, legendary even, was that of Georgia Green, an eighteen-year-old blind music student from Socorro. On July 16, 1945, her brother-in-law was driving her back to the University of New Mexico in Albuquerque after her summer vacation. The blind girl felt the flash and exclaimed, "What's that?" before they drove off the road. The anecdote was widely retold in the press and became the source of speculation: was the light of an atomic bomb so bright that even the blind would see it? Reporter Rolf Sinclair went to meet with Georgia's sister and brother-in-law in 1989. They explained that their sister had enjoyed telling that story; but, although she did not have any vision, she was able to distinguish light and dark so there was no special quality to the atomic light except from its exceptional brightness that she detected. Nonetheless, the legend was born and "the blind girl" became a recurring character in newspapers articles, in novels, and even in General Groves's memoirs.[18]

Testimonies were used a posteriori as tales on the birth of a new age to be used for anniversary commemorations. They were dramatized, sometimes exaggerated, and often tinged with humor or irony. In Socorro, Lee Coker "was slicing bacon at his family's ranch just before dawn when a sudden terrific light and a strange-looking cloud filled the kitchen window. His infirm father, hobbling half-way between the house and the outdoor toilet screamed, 'God Almighty!' and thought the world was coming to an end."[19] The humor, however, obscured the anguish some witnesses felt on that scary day. In Bingham, for instance, Mr. and Mrs. Harold Dean, who ran the post office, thought the Japanese had attacked. In San Antonio, Rowena Baca, granddaughter of Joe Miera, the owner of the Owl Bar & Café where scientists came to drink beer during the Manhattan Project, was shoved under her bed by her grandmother because the older lady thought it was the end of the world. Baca notes, "I think most of us would just like to forget what was a very scary day." The Army engineers camping on the property of her grandfather Joe told him to come outside promising that he "would see something that had never before been seen in the history of the world." Miera saw the light, felt the shock wave, and heard the window panes in his house breaking. His description of the noise was of a mixture of "an airplane, a freight train, and thunder all rolled into one!"

Both the sound and the light impressed witnesses, leading them to believe they were right next to the explosion even though they were miles away. In Ancho, fifty miles and a mountain range away from the Trinity site, Richard Harkey was waiting at the train station with his father. He remembers that "everything suddenly got brighter than daylight. My dad thought for sure the steam locomotive had blown up." Richard comments, "When you see something like that you're so flabbergasted that you just let it go." The expression "in my backyard" was used several times—and would interestingly be used in later years as an antinuclear slogan "NIMBY, Not In My Back Yard." A resident in Carrizozo said, "It sure rocked the ground. You'd have thought it went off *right in your backyard*!" Each had their own interpretation of what they saw and felt. Hugh McSmith thought a plane had crashed in his backyard because he woke to the sound of a B-29 from Kirtland Field that had come out to observe the blast from the air. The noise McSmith heard actually came from the shattering of his water cisterns that had been hit by the shock wave. Another rancher near Alamogordo also had the impression of an airplane crashing in his yard; he said, "It was like somebody turned on a light bulb right in my face!" Roswell Police Chief Lloyd Blakeney confirms the light was "about a mile wide and lit up the western sky for several seconds." Meanwhile, a Navy transportation plane was flying thirty miles east of Albuquerque on its way to the West Coast. When pilot John R. Lugo saw the light, his first impression was "the sun was coming up in the south. What a ball of fire! It was so bright it lit up the cockpit." He then radioed Albuquerque to ask about the fireball. He was given no explanation but was told "Don't fly south!"

These stories told for the fiftieth anniversary of the Trinity Test provided a form of comic relief for an event that was fraught with complex, daunting consequences. Humor is created by emphasizing the discrepancy between the momentous event and the triviality of the scenes in which the witnesses are set or through the reactions of these individuals who had no idea of the meaning of what they saw. At the end of the Oscura range, Bill Gallacher, fifteen years old at the time, recalls his father simply saying, "Damn." He later added, "It was a sort-of-sudden deal especially before you've had your morning coffee." Some of these witnesses were ranchers who had given up their land to the government. One reporter commented humorously on Dolly Onsrud of Oscuro who "woke up, looked out her window and saw a mushroom cloud rising from the other side of the mountains—right about where her cattle-grazing land had been before the U.S. Army took it over three years earlier. She had been none too happy about giving up her 36 sections, and now it looked as if the government was blowing it up."

Some articles also underscored the naïvety of reactions, which can make them comical. William Wrye and his wife, Helen, slept through the event on their ranch twenty miles northeast, but they became aware that something had happened at breakfast when "some soldiers with a black box appeared near the stock tank. 'I went out there and asked what they were doing, and they said they were looking for radioactivity. . . . I told them we didn't even have the radio on.'" According to the article, Wrye's whiskers stopped growing that summer and came back white a few months later before returning to black. Half the coat on his black cat turned white, and his cattle also "sprouted white hair along one side." When these effects of fallout began to show in the summer and fall of 1945, residents tried to get compensation money and claimed they had suffered grievances, which included a newborn "frost-colored 'atomic calf,'" a woman who allegedly lost more than twenty pounds, a man claiming his hair and beard had turned prematurely gray, and a white-spotted black cat in Bingham belonging to Hugh McSmith and renamed "Atomic." A California promoter offered fifty dollars to McSmith to display the cat in a sideshow. In an attempt to alleviate public fear of radiation, Oppenheimer and Groves were photographed at the site within two months of the test. Unaware of the danger, curious residents ventured into the crater. Evelyn Fite Tune, who lived on a ranch twenty-four miles west of Trinity, visited the site with her friends and neighbors soon after the blast. She comments, "We found the hole, we picked up the glass, we climbed the twisted and melted parts of the tower."[20]

Having hosted the birth of the atomic age, New Mexico sought to capitalize on the onset of a new era. The state's participation in the historic battle of laboratories fueled patriotic sentiments and was an opportunity to affirm its American identity. While the national headlines on August 6 and 7 disclosed the dropping of an atomic bomb on Japan, "local media stressed equally the role played by the nearby Manhattan Project facility."

The editor of the *Albuquerque Journal* "noted that the state had played a part in the development of 'one of the greatest scientific advances in history.'"[21] There was also some debate on what would happen to the birthplace of the atomic bomb. Some thought that it might become a tourist destination and that close towns would benefit from it. Others suggested that the site be kept in the same condition to continue research and experiments, but enthusiastic New Mexican boosters had the idea to transform it into a national park. Likewise, some also thought that Los Alamos should be frozen in time as an open museum dedicated to the great scientific achievement. The Laboratory turned out to be much too valuable in the following years to become merely a theme park. As for the Trinity Site, developing and maintaining the military range made the Army reluctant to manage potential atomic tourists, and safety experts were worried that the radiation at Ground Zero might affect visitors and park rangers. The idea did not go through.[22]

First Casualties of the Atomic Era

The Trinity test symbolizes the start of New Mexico's atomic dilemma between benefits and risks, and between pride and fear. The test generated a wave of enthusiasm in the area and the state has tried to ride this wave ever since. However, Trinity also created the first local victims of radioactivity whose existence was long buried in secrecy. Barton C. Hacker, Laboratory historian at the University of California Lawrence Livermore National Laboratory, explains that "safety never commanded topmost concern at Los Alamos. Getting the job done came first. In testing the bomb, however, safety may have ranked even lower than normal." He quotes James Nolan, chief safety planner for Trinity saying, "Possible hazards were not too important in those days. There was a war going on," and Hymer Friedell, second-in-command of the Manhattan District Medical Office, who added, "The idea was to explode the damned thing. . . . We weren't terribly concerned with the radiation."[23]

Even at the highest level of command, priorities did not rank safety higher than secrecy. General Groves listed the principal security requirements in the following order: "barring strangers from the test site, preventing harm to project members, reducing chances that outsiders could learn of the explosion, safeguarding the public from fallout, planning for emergency evacuation, and forestalling any national press reports that might alert Japan."[24] A telling example of the prioritization of success over safety is the concern about the weather. The provisions of experts who predicted terrible conditions were rejected because of the pressure Washington put on Los Alamos to do the test before the start of the Potsdam Conference. On that day, a storm delayed the explosion until 5:30 a.m., and the strong wind blew the radioactive cloud all the way to Indiana (1,300 miles away) where it contaminated the river water used by the Kodak Company. Spotting was observed on films in the fall of 1945, proving, despite the first reports about Trinity proclaiming otherwise, that there had been fallout beyond the detonation zone.[25]

Dr. Stafford Warren, chief medical officer of the MED, was in charge of the fallout operation. Estimating safe radiation dosages (they were as yet unknown) was a complex decision, and one easily adapted to other prerogatives. At Trinity, physicians believed that the dosage limit could be raised to a relatively high level because it was a one-time exposure. The effects of repeated low-exposures would only become worrisome to scientists a decade later. Warren and Louis H. Hempelmann, head of the Los Alamos Health Group, set the limit to a total dose of 75 roentgens (r) over two weeks. At the time the maximum dosage for Project workers, who were exposed on a regular basis, was 1 r/2 weeks.[26] This was based on the dosage that had been set in 1936 of 36.5 r/year. It would drop to 15 in 1950 and to 5 in 1957.[27] Warren's team members, Joseph O. Hirschfelder and John Magee, were responsible for researching the way radioactive particles released by the fireball would spread. They studied the role played by the wind in picking up the particles and "consulted with the top military meteorologists." In their June 16, 1945, memo, the two scientists conclude, "There is a definite danger of dust containing active material and fission products falling on towns near Trinity and necessitating their evacuation."[28] Preliminary tests were run using TNT and plutonium to observe the progression of explosion clouds. These were inconclusive; contamination of nearby streams and communities remained possible. One major concern was the unpredictable effect that rain might have on the radioactive particles.

Warren expanded the safety zone from twelve to thirty miles to encompass the Mescalero Apache Reservation and other smaller towns. An evacuation plan was designed, but the zone also conveniently avoided bigger towns like Roswell that would have been complicated to evacuate.[29] An evacuation detachment stationed at Socorro would be sent to escort families to a hotel in town. In case a whole town was in danger, Major T. O. Palmer, leader of the detachment, had been "provided with enough food and equipment to set up a temporary camp for 450 people. Evacuees would be told that an ammunition dump containing poisonous gas shells had blown up." In the worst-case scenario, "Groves was authorized to declare martial law."[30] The detachment could also ask for help from the Alamogordo Air Base, which could furnish barracks for temporary housing. The evacuation order would be issued as a last resort only if the data collected by monitors confirmed serious danger so as not to jeopardize secrecy over an inconsequential threat.

On the day of the test, Warren deployed forty-four security agents to monitor the habitations downwind. Due to misinterpretation of the wind direction, however, the radiation monitors were deployed upwind, and they did not have time to be relocated. After the explosion, Hirschfelder accomplished his monitoring mission; he remembered driving by a mule that looked paralyzed; then, at a small store, a man told them, "You boys must have been up to something this morning. The sun came up in the west and then went down again." At the Army searchlight post, soldiers were roasting steaks at the time fallout arrived in the form of "small flaky dust particles." The decision was made to bury the steaks and send the crew back to their base camp. The soldiers scrubbed their car, which had turned radioactive; but once the two scientists had returned to the Lab, the radioactivity that emanated from them was still high enough to be picked up by their Geiger counters.[31] In this instance, the monitors arrived on time to tell the soldiers to leave, but that was not the case for all.

The monitors were in charge not only of assessing the level of radioactivity but also of determining whether that level was high enough (dangerous enough) for the surrounding towns and ranches to be evacuated. While personnel at the base knew of the danger, had safety directions to follow, wore badges, and carried dosimeters, the population's safety relied utterly on the monitors' responsiveness and decision making. In the end, the towns of Carrizozo, Bingham, and Vaughan were closest to being evacuated; but the radioactivity levels, while momentarily bothersome, did not prove high enough. The findings of the Los Alamos Historical Document

Retrieval and Assessment Project (LAHDRA) conducted in 2009 led to the following statement about the balance between safety and secrecy:

> Historical records indicate that pressures to maintain secrecy and avoid legal claims led to decisions that would not likely have been made in later tests. Even though exposure rates, total exposures, and alpha counts exceeding pre-established limits were measured and projected; a "cover story" was in place that would have provided an avenue for relatively inconspicuous evacuation of selected residents; and evacuation of personnel, vehicles, shelters, and supplies were on standby, no evacuations of members of the public were conducted.[32]

The one area that raised genuine concern, though, was the place named "hot canyon" by John Magee. Hoot Owl Canyon was twenty miles northeast of Ground Zero in the Chupadera Mesa, right in the path of the cloud, and received the highest intensity of gamma rays. Three hours after the explosion the rates at this location were still extremely high—20 r/h, which was the equivalent of a year's maximum dose in less than two hours. After this discovery, monitors had to take scrubbing showers and get rid of their cars and clothes. They did not return to the canyon until the next day. In the months leading up to the test, the Army and scientists had studied state and county maps and aerial photographs. Palmer's group had made field searches in the countryside to detect, list, and map all habitations in the area. Hoot Owl Canyon was thought to be uninhabited. Yet, on July 17, Hempelmann and Friedell discovered two families.

The Raitliff family was found at their ranch on a homestead parcel: an elderly couple named Minnie and Monroe, their twelve-year-old grandson, dogs, and livestock. The monitors arrived a day after the radioactive cloud reached their ranch, and the personnel decided, after consulting with Warren, to leave them there.[33] The second family was their neighbors, the Wilsons, who ranched up the road. Hempelman calculated that the "Raitliffs received at most a total of 47.0 roentgens [which was below the 75 r limit for evacuation] of whole-body gamma radiation in the two weeks after Trinity." He went to see the families in August and observed that the thick adobe walls must have protected the dwellers from most of the radioactivity. Only Mr. Raitliff had spent time outdoors and told Hempelman he had seen the ground "covered with light snow."[34] The effects of fallout were visible on animals, who had received an estimated 20,000 to 50,000 r on the day of the blast, but not on humans. Milk cows grazing on contaminated ground displayed burns, loss of hairs, and bleeding. Unaware of the danger, the family would then drink the milk and the water from a cistern that collected rain.

Hempelmann reported to the District and requested that "the health of persons in a certain house near Bingham, New Mexico, be discretely investigated."[35] His last visit to the Raitliff ranch occurred in November. The grandson was no longer there, but the couple had a two-year-old niece in their care. All family members looked in good health, and the animals had recovered but still showed some signs of their injuries. Between 1945 and 1947, the Raitliff ranch received seven visits from "LANL and MED medical personnel, health physicists, and Army Intelligence agents, 'under suitable pretext,' to check on the visible condition of the residents."[36] The reason behind the visits was kept from the public and from the families.

Ty Bannerman, a nonfiction writer living in Albuquerque, whose parents came to New Mexico to work at the Los Alamos Lab in the 1950s, wrote about the Raitliffs. He searched for information on what happened to them, wondering why the government did not follow up in the years after the test as the AEC would later do for the Marshall islanders affected by the Castle Bravo blast in 1954. Bannerman's search led him to conclude that "the U.S. government appeared to keep no records on [the Raitliff family's] whereabouts after 1946" and they did not leave any written or oral history to their seven children. Their relatives in Hobbs, New Mexico, did not know about their link to the Trinity story. Bannerman comments, "The Raitliffs were not 'important' to anyone but themselves and their family, few books record their existence. . . . But they were the first to set foot on this path that most of us have lived our entire lives on: that of the unwitting participants, the victims and footnotes to the nuclear age."[37] His essay underlines how many local stories of the nuclear age, evicted homesteaders, desert ranchers, or workers on the Hill have fallen into a historical void. "The history of the bomb is always told through the eyes of scientists and industry . . . We've been left out of the narrative. . . . I hate the fact that we have been treated as insignificant," says Tina Córdova co-founder of the Tularosa Basin Downwinders Consortium who represents the victims of cancers believed to be the result of the test.[38]

It took more than half a century for the health impact of Trinity to be revealed in the form of higher cancer rates in the vicinity. In 1945, however, the only signs of injury were borne by irradiated cattle grazing fifteen to twenty miles downwind from Ground Zero. In December 1945, after threats from influential ranchers to sue over the damage, the Army bought the most severely harmed animals displaying beta burns from those who agreed to sell them. None of the Project personnel had seen the cattle but "the existence of the first atomic bomb casualties was called to [their] attention by a letter dated October 11, 1945, written by a Carrizozo, New Mexico, attorney to the Commanding Officer of the Alamogordo Air Base, filing a claim for injury to cattle of the Red Canyon Sheep Company by the nuclear explosion of July 16, 1945." People from the Los Alamos Scientific Laboratory (LASL) went to verify his contentions, and, indeed, 45 percent of the herd had turned gray on their backs, had lost their hair, or had thick scabbing of the skin. Other ranchers filed claims and other inspections followed suit "with instructions to buy at current market prices all cattle showing visible damage." LASL bought about 150 head of cattle, shipped 17 to Los Alamos and the rest to Oak Ridge and to the Animal Husbandry and Science Department at the University of Tennessee to "see if they developed any long-term effects."[39] It was concluded that the effects of the irradiation were confined to the hair and skin.

The first witnesses and casualties of the atomic age, be they residents or their animals, were thus included in the experiment and provided scientific data about the effects of radiation on living beings. While it could not be said that the local population was used as guinea pigs, contrary to atomic soldiers in the 1950s test series,[40] and although activists in the area later argued they were treated as such, the exceptional circumstances of war, emphasis on secrecy, and medical uncertainties deprived residents of knowing what had happened to them and of being fully monitored. LAHDRA listed the various possibilities through which the public may have been exposed to radiation: directly from the blast, from the cloud passing above, from contamination on the ground, on the skin, hair, or clothing, from inhalation of contaminated air, or from ingestion of contaminated food or water.[41] Information on public exposure to the material that has been published to this day is wanting, especially because internal irradiation was not taken into account. Most ranch houses in the area, like the Raitliff family's, had metal roofs to collect rain water into cisterns for drinking. According to

LAHDRA's conclusions, "internal radiation doses could have posed significant health risks for individuals exposed after the blast" because the detonation of the "Gadget" close to the ground increased the amount and spread of radioactivity, the terrain and wind pattern created "hot spots of radioactive fallout," people living less than twenty miles from Ground Zero were not relocated, and the ranchers' lifestyle necessarily resulted in the intake of contaminated milk, water, homegrown vegetables, and meat from irradiated animals.[42] In July 1945, scientists were primarily concerned with the immediate effects of radiation rather that its stochastic effects—associated with long-term, low-level exposure to radiation such as various forms of cancer and genetic consequences. Hence, "no evidence was found of steps being taken to reduce exposures to ranchers who continued to live in the fallout zone after July 1945." Hempelmann later wrote that some may have been overexposed, but, as there was no way to prove it, it went unnoticed. Owing to the absence of continued monitoring, there is no data available on the full impact of the Trinity test on the New Mexican population.[43]

Meanwhile, another issue arose when it became clear that Trinitite, the green glassy material produced by the first atomic blast, was not a safe souvenir to keep from one's visit to the site. When Martha Bacher Eaton was diagnosed with breast cancer, her doctors were interested in the fact that she used to play with the fragments of radioactive rock that her scientist father had brought home.[44] Soon after the war, the El Rio Motel in Socorro sold the green rocks to tourists over the counter. A Santa Fe bank rewarded new customers with free samples accompanied by a warning: "Do not hold near body more than twenty-four hours." A woman even made a Trinitite necklace.[45] Army officers still display Trinitite at the site for visitors to see today, along with Geiger counters to measure the remaining radioactivity. According to a memorandum by test director Kenneth Bainbridge, the options to cover up the radioactivity were either to seal it with cement or dump it in the Rio Grande.[46] In 1952, the government had the area bulldozed and the Trinitite removed. The Trinity site was considered too small to conduct more nuclear tests, so following experiments took place in Nevada.[47]

Move Closer to "Civilization"?

Like at Trinity, concerns centered on the future of other atomic sites in the aftermath of the war. What should be done with Hanford, Oak Ridge, and Los Alamos? Should they be dismantled or maintained? The outcome of these deliberations would be critical for the state of New Mexico. After 1945, American military and political leaders thought their atomic hegemony would last. Optimists, such as General Groves, thought the Soviets would need twenty years to catch up, whereas pessimists, including some of the Project's most knowledgeable scientists, believed they could do it in four. The latter were proven right as the USSR detonated its first fission bomb on August 29, 1949. The Soviet Union's first atomic test marks the onset of the Cold War—although Gar Alperovitz argued that the Japanese bombings were actually the first event of the Cold War, and others have set the date at Winston Churchill's "Iron Curtain" speech on March 5, 1946, or at the introduction of the Marshall Plan in 1947—but for New Mexico, it gave the cue for swift and impressive changes.

Los Alamos earned international fame and officially became the Los Alamos Scientific Laboratory in 1947. Operation Crossroads in the Pacific provided more work for atomic scientists, as a branch of the expanding research center migrated to Albuquerque and later became the Sandia Laboratory, operating a radical physical transformation on the regional city. The end of the 1940s and the following decade accelerated the development of the state to an unprecedented extent. The Manhattan Project had ignited a local socioeconomic revolution that converted this predominantly pre-industrial agricultural and pastoral region into a symbol of the dawn of a new age at the forefront of technology and scientific research. New Mexican lives were radically changed by the decision to maintain wartime installations in the state. It was the outcome of a long debate between those who desired to make the region a bastion of the emerging national atomic complex and those who rejected the location, ironically, on the same grounds that had motivated its selection in 1942.

The characteristics of the Pajarito Plateau that had made the site so desirable—the seclusion, the topography, the climate, the distance from large cities—went back to being considered liabilities. In the weeks following Japan's surrender on August 14, 1945, General Groves received the recommendation of General Thomas Ferrell to relocate the Laboratory's functions somewhere where top scientists would be willing to stay in peacetime.[48] He suggested Berkeley, Chicago, or somewhere in the East. Ferrell and others thought that the state lacked the appeal to be seen as permanent place of residence for the elite of physics and that, given the choice, talented scientists would rather stay close to the great urban centers, in other words, closer to "civilization." AEC Chairman Glenn T. Seaborg later recalled that some of his colleagues had "maintained that it would never be possible to make Los Alamos attractive for competent scientists. . . . It was too remote from civilization." But "this pastoral setting" did eventually transform into "a modern, bustling community."[49] Satisfaction questionnaires to Lab employees and families revealed that most wanted the Lab to be moved closer to a big city and only 35 percent wanted it to stay in Los Alamos.[50] The sixteen-month interim period until the MED was replaced by the AEC on January 1, 1947, enabled General Groves to exert pressure to maintain the site. He argued that "the United States could never reassemble a similar laboratory . . . except in time of war."[51] War had allowed the District to make exceptional decisions extremely fast, such as gathering up all the worldwide leading nuclear physicists, building a small town in a high desert, but also condemning private and state lands, evicting citizens from their homes, and avoiding lawsuits.

On the other hand, the behavior of the Laboratory staff members aligned with General Ferrell's claim that New Mexico was not adequate to host a permanent research center. The end of 1945 and 1946 was marked by a massive scientific exodus. One of the first top scientists to leave his position and go back to the University of California, Berkeley, was Oppenheimer. Officially, some wished to remain but were committed to other positions, others were "indifferent to the Laboratory's future after victory was won," and others "preferred not to gamble" on the outcome of the discussion.[52] People on the Hill missed the sophisticated and cosmopolitan feel of busy university towns. The romantic charm of the region ceased to operate as the special atmosphere of the Manhattan Project faded. The water shortage during the winter of 1945 did not help the morale of those who had stayed. Complaints about housing multiplied, making it clear that the living conditions would have to be improved to stop the flow of leaving staff and to attract new employees. Preparations for Operation Crossroads and the construction of the fast-neutron nuclear reactor dubbed Clementine helped restart the machine of atomic research, but the desertion of top scientists remained a major challenge for planning operations. In a letter on January 7, 1946, to Vice Admiral W. H. P. Blandy, commander of the naval atomic tests in the Pacific, the new Laboratory director, Norris Bradbury, complained that his scientific personnel were accepting positions in universities and in the industry because the future of the Lab was still uncertain and there was still "one school of thought that said that Los Alamos should be deserted. Put a fence around it, everybody go away, leave it as a monument of man's inhumanity to man."[53]

Bradbury's wish was officially granted on August 1, 1946, when President Truman signed the Atomic Energy Act that created the AEC. To accompany Bradbury's management efforts to quell the exodus, the Army proceeded to make the town more attractive to boost the population's morale and accommodate their suburban needs. Improvements made a significant difference in the town's appeal but further differentiated the Hill from surrounding communities. When the county of Los Alamos was created on June 10, 1949, the division became cartographic. That same year, White Rock, the town's first suburb, was built with laboratory funds as a temporary housing project for construction workers and their families. In the following decades, the scientific community evolved toward normalcy, endeavoring to import the markers of eastern urban life onto the small isolated mesa.

The AEC declared LASL and the Los Alamos community officially permanent in 1948. Once the certainty of the Lab's future had been established and funds had been invested in better living conditions, the trend was reversed, and scientific newcomers poured in once more. The Lab grew proportionally with the infusion of government money and multiplication of contracts to build up the American stockpile. After losing the nuclear monopoly, it became clear to American leaders that the United States had to increase the number of atomic bombs at their disposal. Consequently, the desert became synonymous with seemingly unlimited government funding and astronomical research grants, enticing atomic scientists to participate in this new, scientific conquest of the West. The postwar incoming crowds had little in common with the wartime pioneers who had come in urgency and were not to stay. The incentive and intent were altogether different. The new group were settlers.

New Mexico's metamorphosis occurred through a pattern of colonization involving intensive militarization, seizure of land, and influx of settlers, which was dimmed by the exceptional circumstances of war and the additional dimensions of patriotism and sacrifice. In a region desperate for national recognition beyond clichés of the "Old West," the arrival of new industry was a bonanza. Ironically, it was those very clichés that guided the Manhattan Project toward New Mexico. With new opportunities came the hope for economic prosperity. In the transition from World War II to the Cold War, however, the pattern of colonization was maintained and even reinforced by a mechanism involving federal funds, corporate interests, and secrecy. The Trinity test had allowed a first extension of nuclear science outside the Lab's fence. Connections between centers of scientific research and military bases were repeated, weaving a military-scientific web across the region.

The four years following the end of World War II determined Los Alamos's potential for sustainability. The issues raised by the end of the Project—the scientific exodus, the transition to civilian management, deciding on a research program and purpose, and attracting new personnel—deeply questioned the choice of New Mexico as an appropriate environment for nuclear research. Rather than giving the Pajarito Plateau back to homesteaders and Pueblo Indians, as Oppenheimer allegedly suggested ("Let's give it back to the Indians"), the Army and the AEC made the decision to seek return on the wartime investment by modifying the immediate environment on the mesa to meet scientists' desires. The novelty of Los Alamos gave the town an air of having been transplanted from a suburban community somewhere in the richer part of the country and replanted in a developing society that was still anchored in its agricultural past. It became the "ultimate company town."[54]

As the United States and the Soviet Union entered an armaments race that would redefine the rules of diplomacy and war in the twentieth century, the state of New Mexico became increasingly involved in the construction of America's supremacy. Science, high technology, weapons research, production, testing and storage, uranium mining, and nuclear waste management came to form the new pillars of the state's economy. The financial benefits were enormous for New Mexico and New Mexicans. The multiplication of testing series and the government's request for ever mightier weapons guaranteed constant and rapid growth for the industry while hazards were carefully maintained under a veil of secrecy. The influx of atomic workers to the new research centers had a massive spin-off effect on local businesses and meant that their growing work facilities would provide more jobs for the local workforce. The same colonial framework, in which wartime New Mexico existed and its postwar future exists, has been repeatedly used by the Western World to impose its domination through modernity, science, technology, and the belief in the superiority of its institutions.[55] The particularity of internal colonialism, however, is that the lines between conquerors and conquered are blurred and motivations on both sides are made more complex by personal culture and history.

Six

The Nuclear Golden Goose

From the moment the spotlight hit Los Alamos after the war, New Mexico's economic fate and that of the nuclear weapons complex were tightly interwoven. Many New Mexican families had enlisted members who served overseas during World War II, and the 2,032 war casualties enhanced the state's patriotic gratification in having hosted the Manhattan Project.[1] The people living around the Hill finally had an understanding of all they had been observing for the past two years: the explosions, the soldiers, the gated community, and the influx of Americans and foreigners. Los Alamos and Santa Fe communities formalized their first official meeting in December 1945, when the Hill dwellers "made a grand debut into the society of Santa Fe" at the Museum of Anthropology. A special committee of Santa Fe citizens "replaced [the museum's] Indian exhibits with pictures of atomic experiment and Hiroshima damage."[2] And thus, in one giant leap, the pictures of ancestral techniques and cultures were replaced by those of an atomic future based on new, lethal technology.

The ensuing years were a period of scientific and economic boom, its pace cadenced by Defense Department demands for research and development in the field of nuclear weapons and of applications of nuclear energy. The "Super" program, "a weapon of genocide"[3] according to Oppenheimer, was led by Edward Teller and Stanislaw Ulam and brought back an atmosphere of frenzied intellectual motivation, "a fresh spirit of endeavor" that "enlivened the laboratories."[4] A new technical area was specially built on South Mesa in February 1950, and in three years (1949 to 1952), the population on the plateau almost doubled. The Ivy Mike test on November 1, 1952, brought the hydrogen bomb to the world by vaporizing the atoll of Eniwetok in the Pacific. In New Mexico, the Super program increased funding from $45.4 million in 1951, to $63.4 million in 1952, and $95.3 million in 1953.[5] Military research funds during World War II had averaged $245 million per year but by the first decade of the Cold War, they had reached $1.5 billion, totaling $5.1 billion with indirect research in 1957 (in comparison, the lab budget for fiscal year 2012 was $2.242 billion).[6] The work pace at the Laboratories and the funding levels were in tune with the rhythm of nuclear tests series throughout the 1950s and 1960s.

Less than a year after the Americans, the Soviets also tested a thermonuclear weapon. Strategic thinkers soon concluded that the escalation of armament and technological progress "had made nuclear war obsolete."[7] Out of adherence to doctrines of retaliation, the new doctrine of nuclearism was born. This was understood as the faith in nuclear weapons to maintain national security. In five years, the United States went from an atomic arsenal of two bombs to 298. By 1953, that number had reached 1,161.[8] And the American West became dotted with the new temples of nuclearism: military bases, command centers, proving grounds, training schools, nuclear test sites, bombing ranges, missile fields, arsenals, laboratories, weapons plants, naval yards, and strategic mining sites. The 1950s atomic bonanza primarily affected three places in the central and northern areas of New Mexico: Los Alamos, Albuquerque, and Grants. To the south, the development of White Sands Military Range (WSMR) and the University of New Mexico Tech were other manifestations of the new keen interest in defense-oriented research.

The Beautiful, Modern Town and the "Universal Servant"

"Los Alamos . . . the World's Most Important Small Town" read the title of George Fitzpatrick's article in the August 1949 issue of the *New Mexico Magazine*. His description gave equal emphasis to the scientific achievements of the Lab as to the "amazing" look of the town:

> The town surpasses anything that any high-powered real estate promoter in his most imaginative flights of fancy could have dreamed up. From a collection of tar-paper covered temporary buildings set

in a hurry, Los Alamos is evolving into the most modern community in the West. Housing areas are laid out on far-advanced design for beauty, utility, and protection of playing children. . . . New buildings like the Post Office and the big cafeteria were designed for beauty as well as utility.[9]

The repetition of "beauty" and "utility" combined with the quantification of the town's assets reinforces the impression that the text is a piece of real-estate advertisement praising the aesthetics and practicalities of a brand new suburban area. Modernity, also a central feature in the description, is probably meant to match the company town's new reputation as the usher of a new modern age in a region more often viewed through the lens of its ancient myths. The contrast with Fitzpatrick's preceding article is striking: the reporter had dwelled at length on the geology and volcanic past of the Jémez Mountains for lack of more information on what was happening beyond the fence. He had thus noted the irony of a "laboratory which developed man's most powerful explosive force" being "located on the site of one of nature's most powerful explosions."[10]

Further in the 1949 article, like an agent selling a new vacation resort, Fitzpatrick extols the variety of activities that are available to residents: lectures, exhibits, square dances, sportsmen's movies, rifle shoots, golf tournaments, sketch classes, bowling, skiing, skating, tennis, baseball, hockey, swimming, picnicking, and hiking—a long, non-exhaustive list of attractive pastimes that led some residents to "think that Los Alamos is the most recreational-minded town in the country."[11] Some of the most popular pictures of the Manhattan Project era did feature top scientists on ski outings in the Jemez Mountains. As a result, the tendency to associate the lab's scientific work with outdoor entertainment has remained.

In 1971 Maxine Beckman, who arrived from the University of California, Santa Barbara, accepted a position at the lab because she and her husband "would really like living in the snow area, being able to ski," and she looked forward to the possibility of playing tennis at a facility with sixteen courts. She stressed that while her salary equaled the one she had in California, the cost of living was lower in New Mexico, a much poorer state. She was aware that the government had added recreational facilities on the Hill to "sell Los Alamos to people."[12] The focus on recreation implies that scientists, despite the important work they were conducting, had the luxury of enjoying leisure time. New Mexican workers, on the other hand, had no notion of what leisure time was. When asked about the difference between Los Alamos and his community of Chamita, Rubén Waldo Salazar, who began work at Los Alamos as a lineman in 1944, replied that the main difference was how much money people had and, precisely, that the Los Alamos people "probably [had] more time for recreation."[13]

By 1950 the Hill had completely changed its appearance and looked much like any other permanent American town but with the distinctive feature that most of its population held advanced degrees.[14] A high quality of everyday life and the environment's potential to appeal to sophisticated, city-dwelling scientists was vital for LASL to compete with other employers. The zealous efforts put into the creation of an attractive PhD magnet transformed the Army camp into a unique and thriving boomtown where there were no old people, no cemeteries, no unemployed people, no slums, and no beggars. If someone lost his or her job for any reason, that person would leave and be replaced. Such was the concentration of brains that the community was dubbed "the physicists' Hollywood," where the title Dr. was no longer in use.[15] Unsurprisingly, from the earliest years, the schools were the town's chief asset, praised as "one of the nation's finest school systems with spick-and-span new buildings displaying the latest architecture and furnished with the finest equipment obtainable" as well as "the best teachers and supervisory employees that are available in the country."[16] Maxine Beckman was pleasantly surprised by the fact that the schools, contrary to other places, had no funding problems.[17]

Not all newcomers responded positively to the town's appearance. Harold Gibson, for example, came from Boston in 1975. In Pojoaque, someone told him he would love Los Alamos "if [he was] from back East," but when he visited the town his reaction was: "If I had wanted to live in Connecticut, I would have stayed back East," and he went to settle down in the Valley instead.[18] This account shows that even more than twenty years after the creation of Los Alamos County, because the town had been modeled on eastern standards for beauty, utility, and modernity, it was clearly more at odds with its New Mexican environment than ever. Those who wished to live in an archetypal, tranquil suburban community appreciated the transplanted aspect of the setting, while those who wanted a change of scenery and to discover the cultures of New Mexico preferred settling elsewhere.

The people behind this radical transformation from Army camp to dream town were the New Mexican local workforce hired by the Laboratory's contractors. Two main construction contractors had operated at Los Alamos during the war: M. M. Sundt Construction Company of Las Vegas, New Mexico, and Robert E. McKee Company of Texas. McKee received the Army-Navy Excellence Award for achievements at Los Alamos. He later proclaimed that every one of his employees "was proud" of having participated and "even a small contribution to the entire program was something that no McKee worker would ever forget." At the war's end, the District initiated the creation of a new company headed by McKee and named Zia in reference to the New Mexican flag. The symbol has a religious meaning for the Zia Pueblo, who have been both honored and saddened by the way their sacred sun symbol has been used over the years (e.g., an airline, a school's women athletics, a music band, motorcycles, and portable toilets). The community has asked for recognition and protection of their intellectual property for decades.[19] The Zia Company was not an exception; no one asked the Pueblo for permission to associate the symbol with a nuclear weapons laboratory.

In March 1946, the Zia Company was incorporated and soon became the prime employer in northern New Mexico. The company in charge of management, maintenance, and construction had to hire enormous numbers of blue-collar workers and, as in the wartime years, they were found in the surrounding Hispano and Native American villages. When it began operations in April, Zia employed a force of 1,500 people. These are some of the activities listed in its first institutional advertisement called "What is the Zia company?":

> It is your gas serviceman, your carpenter, your plumber, your electrician, your water, light, and gas company; it furnishes your heat and maid and janitor service for your dorms; it repairs and maintains your streets, your parks, all your official vehicles, and recreational facilities, collects your trash and garbage; it provides cleaning and maintenance service for many of your offices etc.

In sum, there was little that Zia did not do for the Los Alamos residents. It was "the universal servant, everyman's handyman." Laboratory staff members and Zia employees evolved in two separate worlds where one population was at the service of the other. The barrier between the two groups would become a growing issue in later years when the educated children of garbage collectors, plumbers, carpenters, and the like aspired to higher positions, but for these first employees, life changed dramatically thanks to their new jobs. McKee noted in 1950, "Favorable working conditions and pay, and the continued demand for additional labor . . . have beneficially affected the economy of the surrounding area." This rhetoric was readily imbibed by most workers who knew they owed the betterment of their situations to Los Alamos's prosperity. For some time, the company prolonged the bus system, transporting laborers and mechanics back and forth as the Army trucks had done during the war, but the system was substantially reduced by 1950 because practically all workers living off-site used their own cars or publicly operated buses to commute. Some families, like Patricia Trujillo-Oviedo's, had two dwellings; they kept the family house in Chimayó and had a trailer in Los Alamos, leading an urban life during the week and a rural life on weekends. Although Patricia was born in Los Alamos and felt comfortable in both cultures, she considered Chimayó as her home.[20]

The weak competition among job seekers allowed local workers to be hired easily. When in need of labor, Zia would comb the entire Rio Grande Valley for workers. "It is doubtful whether Los Alamos will ever reach a point where construction and expansion will entirely cease, as science is a matter of progressive development," McKee writes confidently. In addition, McKee proudly exhibits a picture of Chief Tafoya, the same Cleto Tafoya, former governor of the Pueblo of Santa Clara, who was pouring soup at the scientists' cafeteria in 1943, in his eagle-feather headdress to illustrate the fact that "Indians of the various tribes form an important part of the workers at Los Alamos" and all of them, "like everyone else, are employed at the same wages and under the same working conditions without discrimination." Emphasizing the slight paternalism of this presentation, McKee affirms that "the building of Los Alamos and its operation have gone a long way toward improving the status of the Indians in the surrounding communities."[21] Some five hundred Native Americans had jobs on the Hill in 1949; they worked as janitors, laborers, and maids. A Los Alamos resident, Mrs. Floyd, was reported saying: "Our Indian maid cheerfully works seven hours a day for $5."[22] An article titled "Boom Town Under Control" underscored the dual effect of Los Alamos "on the ancient routines of these simple communities," which was disruptive but was partially compensated as it "raised their cash incomes by providing jobs."[23]

It is factually demonstrable that Los Alamos improved the economic status of Native American communities, especially if one focuses on the period before its gates were opened to the outside world in 1957.[24] Yet, as early as 1948, Daniel Lang from *The New Yorker* reported that, notwithstanding the financial advantages, "in general the feeling is that it would be better if the Laboratory had never come and the little rich boys [from LARS] had never gone away. The Indians have once again found that the white man's ministrations are not necessarily a boon." These insights came from Edith Warner, who reported what was being said in the Pueblo of San Ildefonso. Pueblo members might have been earning money, but they had lost access to canyons that provided wood, water, and greens for their dances. As the lab grew, local populations were gradually pushed to the margins and became outsiders in their own homeland. "Whenever anything goes wrong," Warner said, "whether it is of a personal nature or general, like the lack of water, they curse the Hill."[25] Her statement clashes with Mrs. Floyd's vision of a "cheerfully" working Indian maid and shows that the relation was solely financial from the locals' point of view as is often the case in a classic, neocolonial family–servant situation.

The Laboratory represented hope for economic stability and a solution to the decline in agricultural activities. Economist Morris Evans explains that the state was "classified in normal times as a surplus farm-labor state," that it had "always been a reservoir of farm workers for the sugar beet fields of Colorado or the sheep camps of Wyoming." The availability of farm work dropped even lower in the postwar years. New Mexican farms were too small to hire laborers.[26] Thus, when the surplus laborers, who had previously gone out-of-state to find jobs in the war industries, such as on the naval shipyards in Oakland, California, came back after being laid off, they turned to the Zia Company for a new livelihood. This was the case of Aaron Martínez's father who decided to go back to Chimayó and found work as a mechanic helper on governmental vehicles at Los Alamos. Martínez wondered whether his father was crazy to come back to "this place" in 1945.[27] In the early years, many jobs in construction, mechanics, and janitorial works on the Hill were seasonal: workers would get hired every summer and laid off every winter. Still, despite the precariousness, the workers expressed genuine contentment with the new situation.

Internalizing Inferiority

The interviews of people who worked in Los Alamos in the lab's early existence (1943–1960) reveal first and foremost a deep sense of satisfaction at the Lab's presence in northern New Mexico. Philosopher Frantz Fanon, specialist of internalized oppression, describes how colonized people have a tendency to internalize the vision of the colonizer and "may eventually feel a sense of gratitude and indebtedness toward the colonizer for civilizing and enlightening the colonized."[28] Both the unprecedented nature of opportunities provided by the scientific conquest and the local populations' plight before the war made them content to gratefully accept a status of inferiority within the new industry. In addition, many internalized the philosophies of nuclear science as they sought prosperity through nuclear jobs. Meanwhile, however, the rise of tensions among community members crystallized around the benefits of working on the Hill and discrimination, which was eclipsed by immense economic advantages, was initially attributed to a form of educational "aristocracy." Not without humor reporter Joe Alex Morris commented in 1948, "The Tech personnel are the *aristocracy* and the oldest residents of the town" because "it is their work alone that perpetuates the subsidized community." They can "get certain favors" and "live on *Snob Hill*, because you can always hire a bookkeeper, but sometimes a nuclear physicist is hard to find."[29] The Hill was dependent on these "scientist-aristocrats" and northern New Mexico became increasingly dependent on the Hill, which prompted interviewees to wonder, "If it was not for Los Alamos, where would we all be? In what kind of shape would we be?"[30]

"Los Alamos has been very good to us," said Bernadette Córdova.[31] Interviewees insist on the potential for improvement, additional training, or further schooling once inside the Lab's system. Technicians gained additional experience on the job as they were trained to operate various machines and heavy equipment. Loyda Martínez tells the story of her father who had a third-grade education. He worked as a janitor and as a truck driver before meeting a scientist, an "Anglo man," who mentored him to become a mechanical technician.[32] But improvement also came from the motivation workers had to escape chronic poverty. Nick Salazar of Chamita wanted to "be somebody" and do something other than work on a farm with sheep or cattle. After his time in the military, where he took administrative courses, he was hired for an administrative job at Los Alamos in 1950. Thanks to his participation in various projects, including the fusion project, and additional schooling in nuclear and mechanical engineering at Santa Barbara, California, and at the University of New Mexico branch in Los Alamos, he attained the highest position for an electrical-mechanical technician in his lab and started conducting experiments on his own. Such treatment ensured that employees supported the Lab even after they retired. Nick continued working as a public relations specialist for the Lab, stressing how it had considerably contributed to the education and expanded the minds of many families in the Valley. His experience on the Hill had led him to see the pursuit of science as "very important" and the work at the national laboratories as "invaluable."[33]

More prosaically, some like Hipólita Fernández, whose husband worked for Zia as a custodian for forty-three years, focused on the comfort they gained thanks to the job they could get easily despite their lack of qualification and education. They did not think of the atomic bomb in terms of deaths or of the impact in the world, but in terms of *jobs*.[34] Alfonso Mascarenes of Dixon said Los Alamos had replaced the rural lifestyle of his community with "good jobs, good homes, good cars," and "a comfortable life" that included more electricity, indoor plumbing, and easier access to water: things he did not have while growing up. He knew "the community in Dixon was not going to do anything for [him]." He wanted a "future for [him]self" and Los Alamos provided that future—he worked there as a chemist for more than twenty-seven years.[35] Oftentimes interviewees mention the exceptional benefits that no other employer in the area would offer (e.g., sick leave, retirement, higher pay). Pedro Martínez speculated that these benefits, like the recreational facilities, originally had been set up for the scientists.[36] The pay—about eighty cents an hour in 1950–1951[37]—was more than in the mines or in the fields, and the work was not as strenuous. Danny Martínez, a mechanical engineer, stressed the fact that he enjoyed an eight-to-five schedule and was not physically tired going home. He said, "I am still having difficulties to believe that this is a job. I was brought up thinking a job was a sweating, tiring activity. It is incredible. I can play football with my son, things that my parents couldn't do because they had to work."[38] The definition of work was thus altered from physically demanding to jobs requiring training and expertise. Along the same lines, the definition of poverty was evolving as well: while the poor farmer of the

1920s could work all day long for meager results, the poor in the industrial world were unemployed. Poverty became associated with idleness.

The before-and-after-dichotomy discourse runs through the majority of testimonies. According to Pedro Martínez, his native community of Alcalde "was dead" before Los Alamos, but people had "really prospered" since. Chimayó was "nonexistent" before, but "now, look at all the beautiful homes there," he said.[39] Josefita Velarde and C. L. Hunter went as far as calling Los Alamos a "life-saver." Hunter added that it was "probably the best thing that ever happened" to the area because Española was "just a sleepy little community." One can note that the adjectives "dead," "sleepy," and "little" echo the vocabulary used by others who described the region before the war through the prism of the old, static western imagery. Hunter's father's dealership was enlarged in 1948 and again in 1954 because sales doubled, even tripled annually. From a business standpoint, there were no drawbacks he could think of because the spin-off effect was dramatic for all who could sell to Los Alamos.[40]

Consequently, working on the Hill became the ultimate goal of local residents and their children. According to José Benito Montoya, working for Los Alamos in 1952 was an honor because "[they] got things [they] never had before."[41] Danny Martínez recalls accompanying his father, a carpenter, to sell wood to the "rich people" who lived up there. One of his uncles was a technician on the Hill and Martínez recollects how they "looked at him with envy" because he had "the constant job," "he had money, everything he wanted," and "it seemed like a really good job to have."[42] Los Alamos became a dream world of which those on the other side of the fence longed to be a part. For Ed Sitzberger, whose father had come to New Mexico in 1920 from Wisconsin, the fascination with Los Alamos dated back to a school football game in the 1940s. He had wanted to work there ever since. Almost protectively, he stresses that it was "always a trouble" for him "to deal with the negative you hear about the Lab." It bothers him "because of the major benefits that the Lab has generated for New Mexico."[43] In many cases, the dream of living and working in Los Alamos became something of a family tradition with all family members working there. Social reproduction intensified the next generation's desire for improvement and social ascension.

Because many in her family worked there, Lucille Sanchez believed that Los Alamos was a "goal" to achieve, too. She took night classes for college credits while in high school and was hired in the visitor liaison office to help foreign workers. On the other hand, Lucille was not sure why she did not get promoted over the eight years she worked at the Lab. She eventually came to the conclusion that she was being discriminated against because other, new people were able to get a supervisory position in the same department, with degrees that were inapplicable to that post.[44] Salaries, even if they were high, were other indicators of discrimination. Rubén Montoya started as a roadman for Zia in 1946 for $125 per month and became a party chief with $200 per month, a sum he alleged to be half what his Anglo colleagues earned. He stresses that he did not pay much attention to the bomb but was more interested in his immediate job, building roads and utilities, which suggests that workers did not have the luxury to consider ethics and politics. Rather, they were satisfied with earning a living in the area even when it compared unfavorably to that of their Anglo colleagues.[45]

Furthermore, employer-employee power relations within the labs may account for the workers' extended silence regarding discriminatory treatment. According to Ramón Frésquez, who started working at Los Alamos in 1945 as a carpenter and then as a clerk, raises (up to seven cents every six or twelve months) depended on recommendations by supervisors who would summon employees in their offices for ratings. This system substantially increased the supervisors' authority over their subordinates, sometimes led to abuses, and emphasized the workers' vulnerability. "They wouldn't ask you if you had anything to say because if you did, they would probably hold it against you later on, so we were afraid," Ramón said. He recalled that one of his white coworkers was promoted immediately instead of receiving a raise and explained that employees did not have any representation at the time; there was no "respect for the lower class" until the 1970s. He was once told that if he "didn't like it," he "had one thing to do and that was to quit," but as he refused, he did not receive his raise that year.[46] With strong competition among would-be employees in the wage system, employers were de facto in a powerful position.

Such situations provided following generations with a powerful motive to pursue an education and reach managerial positions. Charles Montaño, the author of *Los Alamos: Secret Colony, Hidden Truths*, testified that his father's arduous life working with cattle and his third-grade education was Montaño's greatest incentive to go to college. During the summer breaks and after school, Charles had helped his father in construction and thus developed a sense of how hard it was to make a living working outdoors, digging ditches, and mixing concrete. His first contact with Los Alamos was selling manure in potato bags loaded by hand to people who lived there; he could not believe they would buy manure for one dollar a bag. He later attended the New Mexico Highlands University in Las Vegas, declaring, "I wanted to prove to myself I was as good as the Anglos who were living in the nicer homes."[47]

Exposure to abundance, leisure, and sophistication produced a spirit of competition. Aaron Martínez, who was hired by the Lab as a chemist in 1964, said he wanted to be equal to the other chemists. For this reason, he started taking more classes than necessary to complete his degree. He reports his satisfaction in associating with brilliant colleagues, one of whom had been awarded a Lawrence award. When he was laid off in 1977, it came as a shock, but degrees had become as much a criterion for recruitment as for dismissal. He believes, "I didn't have a Ph.D. and other people who got fired didn't have a Ph.D. either. So, I thought it was that degree, no matter how productive or good you are, you're protected."[48] For a long time, employees believed their inferiority in terms of education rather than their status as minorities was the source of discrimination at the labs, a discourse readily supported by other employees and officials.

Envious feelings were due to the increasingly stark contrast between Los Alamos and the workers' communities. Richard Cook commented that Española in the 1950s had nothing to compare to the swimming pools, the new schools, the recreational centers, and other facilities at Los Alamos. The town could not afford such facilities. Statistics by the Bureau of Business Research at the University of New Mexico in 1953 confirm the yawning cultural and economic gap between the two areas:

> The new county still lives almost entirely on Federal Government funds and has no roots in the State's traditional farming and ranching, mining, lumbering, and trade. . . . In nearly every comparison of the 32 counties by income, employment, housing, health, and similar characteristics, Los Alamos stands either first or last. It leads in such favorable indicators as income per capita, percentage of college graduates, and number of homes with electricity; it is lowest in such attributes of substandard economy as inadequate housing, death rates, and unemployment.[49]

Jealousies toward people who had access to the Hill increased in the following decades as fewer positions were available. The positive aspirations inspired by predecessors were replaced by tensions among members of the same age group between those who left and those who were left behind. Francisco Leroy Pacheco addresses this issue when talking about being considered as an outsider in his home community. He recalls the terrifying change it was for a four-year-old boy to move from Mora to Los Alamos, which still looked like a military base with security guards. When his father began work for Zia in 1949, houses were available only for scientists and engineers. The company eventually built small apartments for workers in the early 1950s, but the total nonwhite population in the town then was fewer than forty people.[50] He recalls how rich people were, how well they ate, and how well they were dressed. School was a culture shock. He sometimes felt out of place because he had only two shirts, while other children wore a different shirt every day.

Francisco Pacheco explains that the first time his father applied to Los Alamos, about two hundred people from Española, Taos, Pojoaque, Las Lunes, and other communities were waiting in line to turn in their applications. He believes his father was hired because he was talented with his hands, could speak English well, and had a light complexion. After working for Zia, he worked for the Los Alamos branch of the University of New Mexico as a custodian. Some of his father's friends, who had stayed in Mora, became jealous of them. When they came back on weekends, "people would say the Pachecos from Los Alamos are in town." They believed living on the Hill meant they had everything they wanted, but they "did not live in a mansion." His brother and he were considered outsiders, "branded" because they had left. As a result of these tensions, there were a few violent altercations. Francisco also mentioned one element, which was an important vehicle of social and spatial stratification in Los Alamos: the points system established after the war to control housing. This system meant that

> People in the same trade would live in the same neighborhoods at that time. When the government sold the houses up there, you got points for the job that you had (the higher job you had, the more points you got). I asked my dad once "how come we don't live there, where Jim or James live?" He would say, "well, you know Jim's dad is a scientist, he has 5,000 points, and I only have 1,000." So

really quick, it became the blue collar/white collar separation of Los Alamos. In our neighborhood, you didn't have a single scientist. A couple of people who worked for the Lab, maybe, but all the others worked for Zia as electricians, plumbers, roofers, etc. Economic discrimination, that's what I call it. . . . Then, they had 2 Junior high schools: one became the blue collar school and the other the white collar school. They had the new gym, the new lockers, the new books. At Pueblo Junior High, we had the old versions of all that.

Competition intensified when the town opened its fences in 1957. While in junior high school, girls showed disappointment when Francisco told them that his father worked for Zia. He heard for the first time the term "Mexican" or "Greaser" from children talking about his friends who lived in Chama, Tres Ritos, Piñasco, and Española.[51] So while the youth were increasingly estranged from their native communities, they also faced rejection (housing) and segregation (schools) at Los Alamos. This created an in-between situation as has often been observed when a culture of wealth develops next to poorer areas: those who manage to leave one sphere to enter the other become double outsiders, that is, both in their native and in their adoptive communities.

Bomb Assembly In the Duke City

Job opportunities were not limited to Los Alamos in the postwar decades. Before the end of the war, in March 1945, the attention of the Manhattan Project's leaders had turned to the "Duke City" to conduct Project A (for Alberta), which consisted in making sure that the bomb would be a practical airborne military weapon. The closest large airport to Los Alamos was Kirtland Field—formerly Oxnard Field. A group was created in Los Alamos, in July 1945, to handle future weapons-development engineering and bomb assembly under the leadership of Dr. Jerrold Zacharias from the MIT Radiation Laboratory. Zacharias' group, named the Z Division after him, was moved to Albuquerque in September. Z division was to become the Sandia National Laboratories, a direct descendant of the Manhattan Project.

In the aftermath of the Japanese bombings, military leaders worried that stockpiling was not going fast enough. General Nichols proposed using outside contractors to build most bomb components and transferring the final assembly work to a special technical military unit. Los Alamos could thus be relieved of its production responsibilities and focus instead on the development of new bomb types. The Z Division was deemed perfect for the job. In order to build a stockpile of the Fat Man model, the division would use the designs produced by the X (explosive) and G (gadget) divisions, develop models, and conduct surveillance tests. Sandia Base would therefore need a testing range. In September 1945, representatives of Z Division selected Range S-1, approximately twenty-five miles southwest of Albuquerque. Part of the range, located west of Los Lunas and east of the Rio Puerco, rested on land belonging to Isleta Pueblo. Arrangements were made by Glenn Fowler, head of Sandia's testing group, with an official of the Pueblo for permission to use the area. Through this transaction, more Native American land was integrated in the weapons production complex—more Native land because the San Ildefonso Pueblo had claimed their right to LANL land on the grounds that it belonged to their tribe and they had sacred sites on the mesa.

In mid-1946, General Groves sent a special Army battalion to Sandia Base to be trained and represent the military for surveillance, field tests, and weapons assembly. This group of sixty-three junior Army officers became known as the "Sandia Pioneers," again attuning atomic science to the nation's explorative spirit. Since the general believed "that nobody over thirty-five would have a mind flexible enough to understand nuclear physics," most group members were well under the age of thirty. At the same time, military personnel stationed at Tinian were transferred to the Walker AFB in Roswell, conveniently close to Kirtland, so that pilots and B-29 aircraft would be available to provide air support for testing operations. At a rate of two tests per week in 1946, all bomb instrumentation completed in Los Alamos was sent to Sandia for final checkout. The unit would be loaded in a B-29 at Kirtland, and the Z Division convoy would proceed to the Los Lunas Range for testing.[52]

The intensification of tests that year was due to operation Crossroads in the Pacific. On that occasion, Senator Carl Hatch, a member of President Truman's Evaluation Commission, chalked "Made in New Mexico" on the side of the bomb's canvas security cover "to make sure that New Mexico personnel received just recognition for their substantial technical contribution."[53] Finally, in the fall of 1947, the AEC transferred all remaining activities of weapons assembly from LASL to Sandia, enabling the original weapons laboratory to focus on research. Thus, Albuquerque's Sandia Laboratories, seventy-five miles south of Los Alamos, became the world's first assembly line to produce nuclear weapons. The connections among Sandia, Los Alamos, and Air Force bases formed the skeleton of a well-oiled mechanism that generated immense economic activity in the state. In the exchanges and collaborations among the sites, the cogs of the nascent military, industrial, scientific, and academic complex were at work. The growth of this complex tightened the bonds between the facilities that pre- and postwar militarization of the area had put on the map of New Mexico.

Two years later, the Regents Committee on AEC projects stated that Sandia's work of weapons production, stockpiling, and surveillance, especially in peacetime, was not appropriate for University management. A transfer to corporate management was requested. Paul J. Larsen, director of the Sandia Branch since 1947, proposed to the AEC that the Laboratory be operated as a New Mexico state corporation to be known as Sandia Laboratory.[54] The AEC, however, favored an impartial manager to run the branch, so the Western Electric Company and the Bell Telephone Laboratories (BTL), which both belonged to the American Telephone and Telegraph industrial complex (AT&T), accepted operation of Sandia. BTL would be responsible for research and development (R&D) while Western Electric would handle the manufacturing.[55] The transition was a complicated move because of the growing demand for weapons in the 1950s. The persistent deterioration of global politics intensified production and added the task of storing the stockpile. In 1956, the position of vice president of Research and Development Technical Services was created to take charge of storage in the foothills of the Manzano Mountains, just east of the Sandia Base. Originally called Site Able and later renamed Manzano Base, it was one of the six original National Stockpile Sites. Manzano and Sandia Base later merged with Kirtland AFB in 1971. Thus, by the end of the 1950s, nuclear weapons production in New Mexico went from design all the way to include storage.

The rapid growth in demand for Sandia products and services caused many difficulties, and recruitment issues in particular. At its beginning, the force counted only 370 people, but after the announcement, in April 1948, that Sandia would become an independent branch of LASL, personnel grew to 470. By the fall of that year, the figure had exceeded 1,000 employees.[56]

The recruitment campaign led by the head of personnel brought in new employees at the rate of 25 per week in 1948. The Washington AEC office helped advertise job queries across the country in order to locate the talent necessary to work at an atomic weapons factory. This was no easy task because talents and qualifications in cross sections of industry were needed, but also because the Laboratory had, once again, to "sell" New Mexico to potential recruits. In parallel, the facility also grew physically: between the formation of the corporation in 1949 and 1963, the land occupied by the Lab at Sandia Base grew from 458,000 to 1,651,000 square feet.[57] However, while the Sandia Corporation continued to spread, it did not own any of the land because its headquarters were situated on a military reservation on land ceded by the state of New Mexico to the federal government, making all buildings government property. Nonetheless, Sandia was under civilian control represented by the Santa Fe Operations Office, which became the Albuquerque Operations Office (ALO) of the AEC in 1956. This intricate status was coupled with the corporation's confused identity due to its diversified activities: should it be a manufacturing organization, an R&D institution, or a field test branch? Research focused on issues such as the height at which an atomic bomb should explode to maximize damage, loading systems to board weapons on aircrafts, the shape of bombs and their casing, ballistic systems, and guided missiles. Gradually, these activities gained ground through organizational changes that decreased its production role while augmenting R&D and testing activities.

The intricacy of ownership and control among the industry, the government, and the Army in Albuquerque reflects the organizational complexity of the nation's nuclear weapons complex as a whole. What role was the state of New Mexico to play in this machinery? It was to participate in providing the means of production and local support for national policies. The state furnished land and labor for the base to expand. Personnel needs both at the managing offices in town and at the corporation meant there were jobs available at all technical levels for scientist transfers and for local skilled and unskilled workers. The majority of workers hired locally were technicians, security guards, secretaries, clerks, construction workers, janitors, and handymen. Furthermore, the Zia Company also participated with the construction of permanent buildings at Sandia.[58] The state thus saw the corporation as an economic asset to create partnerships with public and private institutions. This remained an important political strategy as in 2003, for instance, Governor Bill Richardson announced a partnership between SNL and an Albuquerque investment fund to create up to two thousand new jobs in New Mexico over four years.[59]

In return for its investment in land and manpower, the state received economic gains from Sandia's purchasing activities. Jobs, dollars, and people fed into the creation and prosperity of other businesses and higher education institutions. The corporation bought components and materials throughout the country with suppliers in forty-one states, but many of its purchases were made locally. In 1951, Sandia's first multimillion-dollar purchasing year, orders placed with New Mexico businesses totaled $3,476,821.[60] A 1958 article about Sandia's success recounted its humble beginnings: "At the close of World War II residents of Albuquerque wouldn't have given a plugged nickel for the future of Sandia Base. Thirty military personnel with a handful of civilians had the job of cutting up junked aircraft for the scrap heap."[61] By then, however, the Sandia Laboratories on the civilian side, and the Armed Forces Special Weapons Project on the military side, which was the main "customer for the AEC's ordnance products," pumped an annual payroll of $75 million into the New Mexican economy.

The "wonderful things . . . being done" at Sandia stimulated optimism for the future of New Mexico's economy and led newsmen to write enthusiastically about the prospects heralded by the Los Alamos–Sandia alliance:

> Electronics will play a major part in shaping New Mexico's industrial future, and research at Sandia is hastening the day when private industrial development will take its place alongside government contracts as an important factor in the State's economy. Aside from Albuquerque's boom, the important thing about Sandia and its mother-project, Los Alamos, is that they have given to New Mexico a ground-floor position in the industrial structure of the Atomic Age. Whatever happens to the weapons program, Sandia's $60 million laboratory and highly trained technical and scientific personnel would be the foundation of a new and ever-increasing economy for this industry-hungry state.[62]

The longing for stability appears clearly in the above paragraph that underscores the kinship between sites and the importance of developing skills that would be transferable to other fields, were the nuclear branch to collapse. Comparing the state to a hungry child who has found a new caregiver indicates that the hardships of the past were fresh in memories and nourished high hopes for the future of nuclear science.

Both the economic and demographic growths of Albuquerque were exceptional in the late 1940s and 1950s. While the state's rural population grew by approximately 11,000 between 1940 and 1950, its urban population grew by close to 150,000.[63] People arrived in waves from other states but also from other New Mexican counties in the hope of being hired at the Sandia Corporation or Kirtland Base. The military and weapons industry replaced the railroad as the city's main source of employment. By 1957, about 30 percent of Albuquerque's labor force worked either at Kirtland or Sandia. The median family income in Bernalillo County had doubled in a decade.[64] In 1960, forty-three thousand college graduates lived in the city and new homeowners (75,374) largely outnumbered longtime homeowners (12,037).[65] However, while the city's population of about fifty thousand in 1945 quadrupled in fifteen years, the percentage of Hispanos dropped from 35 percent to 20 percent because of the influx of highly qualified Anglo "atomic immigrants" from all over the country.[66] Marc Simmons explains that although "unskilled and displaced poor, many of them Hispano rural folk . . . abandoned their subsistence farms in droves to seek employment in Albuquerque's burgeoning job market," their exodus from rural to urban areas did not prevent their declining within the total municipal population "in the face of a mounting inflow of Anglos."[67]

The atomic transplants from other states generally were a young, sophisticated, and affluent population. They were recruited on the basis of their education level and soon constituted the elite at Sandia, which employed 1,649 degree-holders in 1957. The newcomers—white-collar, upper-middle class, and well-paid—and their families settled in the heights of the East Mesa, which became known as the Albuquerque Heights or the "science suburbs" on the outskirts of the city. As early as during the war years, the influx of residents had posed a challenge because of shortages in residential housing. However, postwar growth encouraged developers and city boosters to build frenetically. The number of building permits skyrocketed with the expansion of Sandia. In 1947, the city created a Building Department to help process all requests. Three years later, "an estimated 362 residential business builders and subcontractors operated" in the city;

"they employed almost 6,000 skilled and unskilled workers with payrolls exceeding $20 million per year."[68] Massive construction thus contributed to the economic boom.

Albuquerque's phenomenal demographic growth enabled the city to achieve a metropolitan status and become a major urban hub in the southwest, but uncontrolled expansion was also characterized by an aggressive market for prospectors and drew new boundaries within the city. The urban landscape became increasingly polarized between the popular, working-class, mostly Spanish-speaking neighborhoods of the South Valley and the East Heights. Polarization also manifested itself in politics with a Republican stronghold in the Heights and predominantly Democratic voters in the valley.[69] The arbitrary redefinition of the Duke City in the twenty years that followed the war foreshadowed some of the socioeconomic difficulties that the state would confront once the first signs of backlash from its military, industrial, scientific, and academic complex began to show.

Rockets and Yellowcake

New Mexico's laboratories quickly grew and increasingly impacted their environment. They were the backbone of New Mexico's economic new deal. However, other installations furthered the development of the nuclear industry. To the south, the steady development of White Sands Proving Grounds (WSPG) was a similar boon for its surrounding communities. Colonel Harold Turner, first commander of WSPG, "prophesized that this state [would] be proud of the work accomplished within its borders."[70] Alamogordo, the town closest to WSPG, had been founded in 1898 by Charles B. Eddy as the terminus of a railroad line that would tap the logging areas of the Sacramentos. The M. R. Prestige Lumber Company, with a payroll of about three hundred, was the biggest industry in the prewar years. After the war, Alamogordo, like Los Alamos, was associated with the bomb that had been tested on the adjacent bombing range and "cashed in on the resultant publicity." Tourists would stop there "to have the atomic bomb crater pointed out to them or ask directions for reaching the crater."[71]

The cooperation between Los Alamos and White Sands was extended when weapons research turned to long-range rockets. As early as 1930, Dr. Robert Goddard had worked on rockets at Eden Valley near Roswell. From the start, the Alamogordo Army Air Field was informally considered as a guided-missile development site, but it was officially selected in 1944 as a US Army Ordnance Corps test site. During the liberation of Europe, the American Army captured one hundred German rockets. The V-2, or "Vengeance Weapons 2," rockets had been under development since 1936 through the efforts of scientists led by Wernher von Braun. In May 1945, the first components were secretly shipped to New Mexico and the first captured V-2 was tested at White Sands on April 16, 1946, under the supervision of German engineers and scientists, including von Braun, who had come to the United States through Operation Paperclip.[72] So, after gaining fame from its association with the atomic bomb, Alamogordo soon became known for being the town where you could hear rockets being fired, even though rocket launches remained hazardous. Journalist Orren Beaty reported, "One of the latest launchings frightened spectators more than any previous malfunction, as the rocket climbing about four times its length into the air, spun to the east, then wobbled back to the west, in the direction of newspaper writers, Army, Navy and civilian observers."[73]

In 1948, a local café "erected a neon-lighted replica of a jet-propelled rocket as an advertising sign." The *Alamogordo News* quoted a Navy officer from WSPG saying that as many as ten thousand people were employed at a California proving ground, but this number would be "peanuts when the White Sands will come into its own."[74] Optimism and hope for future prosperity was widespread. The only difficulties were shortages of water and housing. Soon, wells were drilled, investigations for new springs in the mountains were conducted, and plans were made to finance pipe lines. The site of White Sands itself became a boomtown mixing permanent and temporary units. By 1954, the desert outpost had become "the principal Ordnance Corps installation for the execution of all technical and engineering responsibilities associated with testing guided missiles and rockets."[75] The site had it all: a school, a five-hundred-seat theater, an enlisted men's service club, bowling, baseball, tennis, handball, ping-pong, a library, a chapel, a swimming pool, and an auto hobby shop. The housing, however, remained so inadequate that many civilians and military families commuted every day from Las Cruces or El Paso. As a result, Las Cruces, twenty-eight miles west of the installation, was the second city to collect substantial profits from the presence of WSPG. A 1954 report "placed the annual payroll of Las Cruces residents at WSPG at $10,000 a day," and the total payroll for the post, both civilian and military, was upward of a million dollars a month.[76]

In parallel, the site represented opportunities for local students through programs such as the White Sands Student Trainee Program, a joint educational venture between WSPG and the Agriculture and Mechanical Arts College (A&M), renamed New Mexico State University (NMSU) in 1960. In 1952, President of A&M Dr. John W. Branson and Brigadier General G. C. Eddy, commanding general of WSPG, met to discuss the possibility of a partnership. White Sands needed trained engineers and technicians while the college, "its enrollment lagging because of the Korean War, had facilities for more technical students than it was able to obtain."[77] This example illustrates a wider trend in New Mexico's education institutions as change was introduced by the regional scientific and economic boom. The University of New Mexico redirected some of its focus to physics and science and opened a branch in Los Alamos. The most telling example, however, is probably the New Mexico School of Mines (NMSM) in Socorro. In 1893, when Socorro was a mining boom town, NMSM opened with a focus on chemistry and metallurgy to train young mining engineers. With the return of veterans after World War II, the school's enrollment soared, and its president, E. J. Workman, a physicist, changed its focus. He had worked on weapons development during the war and brought a new group to the school, the Terminal Effects Research and Analysis group (TERA), to work on weapons testing and analysis. In 1951, the college's name became the New Mexico Institute of Mining and Technology, or New Mexico Tech.[78]

The various programs at White Sands, such as the Apollo and space shuttle programs, and contracts with other universities helped New Mexico Tech and New Mexico State become major research institutions. Furthermore, such instances of cooperation and partnerships between military-scientific facilities and academic institutions meant that the nuclear weapons complex was not just a scientific, military, and industrial structure but also an academic one.

With each new site, atomic newcomers conquered entire neighborhoods or settled in a brand-new town. Nowhere, however, is the analogy between the scientific conquest and westward immigration better illustrated than in Grants, the soon-to-be "uranium capital of the world." As Los Alamos, Albuquerque, Alamogordo, and Las Cruces expanded due to their proximity to atomic weapons facilities, military reservations, and test fields, the region of Grants in the northwestern part of the state was similarly transformed after the discovery of uranium ore in the vicinity. New

Mexico's nuclear industry became a cradle-to-grave economy, starting with the mining of raw material and ending with the management and storage of aging nuclear weapons.

The little town, seventy-eight miles west of Albuquerque, founded in 1882 when the Santa Fe railroad reached the area, was chiefly supported by lumbering, mining, and ranching until 1940 when "some Arizona vegetable growers discovered that the volcanic soil of the Bluewater Valley just west of Grants was ideal for raising beans, peas and carrots."[79] Thanks to the carrot industry, the town doubled its size between 1940 and 1950; it then proclaimed itself the "Carrot Capital of the World." In 1950, a Navajo Indian named Paddy Martínez found a piece of Todilto limestone containing the yellow mineral uranium at Haystack Butte on the lands of the Santa Fe Railroad. After this discovery, Grants became a magnet for prospectors and mining companies. Journalist Wayne Winters wrote in 1951, "Uranium has come to Grants! And just in the nick of time, too, for this 68-year-old northwestern New Mexico town was starting to feel the pinch, following the petering-out of some of its industries."[80] The deposits were mostly found on Santa Fe Railroad lands, but also on the private lands of a few ranchers who held mineral rights to their finds, and on public lands where individual prospectors competed to stake their claims. A rich deposit was found on Mount Taylor. The news was particularly welcome there because the New Mexico Timber Company had closed down its local plant in 1951 and some seventy-five heads of families had lost their jobs. The pumice mine south of Mount Taylor was no longer working on a large scale either. So, like in other areas of the state struck by poverty and unemployment, the arrival of a promising industry was viewed as a miracle and "the local populace . . . became jubilant."[81]

With the escalation of weapons production, uranium became one of the most important materials in the world. Fortunes could be made in the Grants Mineral Belt. The yellowcake rush attracted all kinds of prospectors, including young couples who all "came to Grants for the same reason: to make money."[82] Prospecting handbooks excited these novices' hopes to strike it rich without major investment of capital, time, or knowledge. These migrants were not much different from the fur trappers, Forty-Niners, or cowboys of the previous generations. According to journalist Kevin Fernlund, "Viewing the activities that occurred on the Colorado Plateau at mid-century in terms of the Old West provided the nation with a familiar, if mythic, counterpoint to the frightening realities of a world on the brink of nuclear devastation."[83] Although the uranium rush was not exactly another example of rugged American individualism in that it was a government promoted, supported, and controlled mineral rush, it was nonetheless a conquering move.

In the wake of Martínez's discovery, the Anaconda Copper Mining Company found a large uranium deposit on the Laguna Pueblo Reservation east of Grants. The company had to reach an agreement with the nation before it could begin mining. According to specialist Michael A. Amundson, "The question of whether the Lagunas would comply constituted a classic pattern in Native American history. Until a precious commodity was discovered on the lands, the government considered most Native lands worthless. But when uranium was discovered, the country's national defense was suddenly at stake."[84] In late 1951, Anaconda signed a contract with the AEC to build a uranium-processing mill at Bluewater to process ores from the Jackpile mine on Laguna land.

"Uranium Discovery Makes Laguna Indians Modern 'Rags to Riches' Story" was the title of an article in July 1957, which recounted how the community of approximately 3,600 members had seen its income rise from less than $1,000 in 1953 to nearly "$3 million salted away and a yearly income of more than $1 million." As stated in the lease, the Anaconda Company also employed Laguna men and trained them in the use of heavy equipment. As a result of the royalties from production of uranium at the Jackpile mine on the reservation, "homes are being modernized . . . the villages have electricity . . . and TV antennas dot the rooftops. . . . Trucks, automobiles, TV and radio sets, washing machines and electric stoves are becoming more common at the pueblo now. A few years ago there was no electricity and only horses and wagons."[85] Modernity and progress, measured in the acquisition of technology, transport, and appliances, were again presented as prime instances of benefaction to local communities.

In March 1955, Lewis Lothman from Houston found another substantial deposit near Ambrosia Lake and four other yellowcake-processing mills were built there—the Homestake mill, the Shiprock mill (1954), the Kerr-McGee mill (1957–1958), and the Phillips Petroleum mill. The Kerr-McGee, Phillips, American Metal, and Homestake companies all signed contracts with the AEC in 1957. They worked with smaller companies to dig the mines and build the mills, employing more than four hundred people. The stream of people coming to Grants transformed the community of farmers into a uranium boomtown. The number of residents rose from 2,251 in 1950 to 10,274 in 1960. Uranium was everywhere, including in the local culture: in 1956, a local festival named the Uranium Prospector of the Year; a beauty pageant elected a Miss Atomic Energy in 1955, whose prize was a truckload of uranium ore; and a "Uranium Café" opened for business on the legendary Route 66 that crossed the town.[86]

Thus, throughout New Mexico, the Manhattan Project and its resulting offspring revolutionized towns and communities. New opportunities propelled the state from its pre-industrial past into the future: a nuclear age. New Mexico's identity in the nation became associated with atomic "firsts," from the world's atomic weapons laboratory, the first atomic explosion, and the first assembly line for building atomic bombs, to the first burial site for low- and medium-level nuclear wastes more than fifty years later. This observation led historian Ferenc Szasz to write, "The story of modern New Mexico has ever been intertwined with the emergence of atomic America."[87] The interweaving and simultaneity of the two emergences (modern New Mexico and atomic America) created dependence: America needed the production of more nuclear devices to wage the Cold War and New Mexico needed America to continue fueling its economy with more contracts.

Seven

A Federal Sponsor

The war remodeled the American West through diversification and intensification processes. Government interventionism spurred the reconstruction of economies that had had havoc wreaked upon them by the Depression. According to the "Nash thesis," the postwar boom was accompanied by the West's emancipation from the previous system of economic dependence on eastern states thanks to the federal government, which "promoted the restructuring of a natural resource-based colonial economy into a technologically oriented and service economy stimulated by massive federal expenditures."[1] In the New Deal era,[2] federal intervention had concentrated on land management, soil restoration, and on alleviating the effects of poverty through the Works Progress Administration (WPA), the Civilian Conservation Corps (CCC), and the Social Security Act of 1935. After the war, however, it was meant to boost new economies by becoming their major source of capital and demand. Immediately following victory, because of the demobilization of soldiers, there were fears that depression might reappear, hence the Employment Act of 1946. Westerners counted on the infused capital to stimulate growth and gain autonomy. In fact, their reliance on federal aid demonstrates one of the greatest paradoxes of the West that contrasts the myth of free, rugged individualism with high federal involvement throughout its history.[3]

World War II propelled government action into the fields of technology, science, and engineering, a combination that became a beacon of hope for economic self-sufficiency to combine with or replace primary-sector industries. In the Schumpeterian interpretation of Kondratieff cycles, technological innovation is "the main driving force of change giving rise to the growth and decline of economies."[4] The atomic bomb triggered such a wave in New Mexico and the flow of federal money served to sustain the mechanism with a classic multiplier effect. Federal spending in the 1940s "fueled urban growth in the state to such an extent that Albuquerque became known locally as 'Little Washington.'"[5] The federal government was at the center of a new economic system of partnerships among people in the private and public sectors, including government and military officials, corporate executives, contractors and suppliers, legislators, research organizations, university scientists, and workers in defense plants. The partnership system became known as the military-industrial complex. The boundaries between branches became increasingly more difficult to discern. One consequence of this conglomeration was that private companies of all sizes depended on the government for funds, while the government depended on private businesses for technology and innovation, the two major battlefields of the Cold War. Meanwhile, federal policies inspired by "big government" tended to favor big businesses and corporations that had their headquarters in the East—New Mexico's uranium industry and the creation of the Sandia Corporation illustrate this tendency. The availability of jobs requiring skills within the big government–big corporation system stimulated the inflow of talented professionals and quickly reinvigorated Depression-struck economies.

Although Nash's "federal landscape" thesis suitably supports the better side of New Mexico's postwar economic revolution (the nuclear golden goose), the purpose of this book is to take into account the long-term impacts of the nuclear weapons complex and, consequently, those of the federal government. Notwithstanding New Mexico's unequivocally formidable economic growth, the overwhelming part played by the federal government and big corporations did not translate into substantial independence. The arrival of atomic science was actually another cycle of conquest[6] that transferred New Mexico's dependence from eastern markets to the government and the nuclear weapons complex. While before the West needed eastern markets to sell off its resources, it now needed its main buyer, the US government, to sell off the weapons (and other goods) produced by the nuclear complex, which also relied on federal investment to continue production. Supply and demand were thus both created and determined by the same agent. This imparted immense control over the fate of local economies to the national government as it was present on both ends of the production chain. As a result, economic self-sufficiency was not achieved. Rather, the region was facing another kind of dependence at the mercy of changes of direction in foreign affairs.

Federal intervention was a "familiar channel" through which postwar changes flowed. Richard White writes, "To many Westerners it seemed that the West had merely switched masters. That liberation from eastern capital meant only that the Federal Government exerted an even greater power over the destiny of the region." Federal control over western states already rested on its great landholdings in the region, on the weakness of western political parties, and on the region's past as an internal colony. At the end of the Cold War, despite this federal sponsoring, the region was still characterized by "an extractive economy in a world market; an economy plagued with excessive competition and a shortage of capital; an economy dependent on government aid, outside capital, and outside expertise."[7] The war and federal intervention thus revolutionized the West while, at the same time, reviving its myths, perpetuating its traditional uses, and prolonging some of its struggles. Likewise, New Mexico's postwar evolution is ambivalent for it associates dramatic change with continuity. The metamorphosis was spectacular enough to be interpreted as the emergence of a novel economic model that would eradicate previous difficulties when, in fact, federal capital, science, and technology did not break the patterns of dependence that had continuously plagued the state's economy and its populations. Old issues remained and new ones arose.

Uncle Sam Buys Uranium

Undoubtedly, the most telling example of the dependent relation between New Mexico and the federal government—uranium mining—is also an illustration of how corporate and government agents interacted within the nuclear weapons complex. Uranium mining took the same pattern as other mining cycles, from small individual claims to large corporate domination. As pre–World War II extracting industries had been symptomatic of the state's colonial economy, uranium extraction between 1950 and 1990 reflects how outside forces still controlled its new industries. Until 1970,

Uncle Sam was the only legal buyer for uranium in the country. By the time the system changed, the uranium industry in northwestern New Mexico was barely surviving. According to Michael Amundson, "The history of the yellowcake communities is also a study in the intricacies of economic colonialism." This type of colonialism, which combines corporate and government interests, is the link between local and global economies as it exists in many other areas of the world, where local businesses and workers pay the price to access large, distant markets. Amundson comments, "Throughout the history of uranium production, outside forces controlled the fate of the yellowcake towns. The most obvious agent was the Federal Government."[8]

The first uranium boom in Grants between 1946 and 1958 was characterized by the monopsony between private producers and the federal government established by the Atomic Energy Act of 1946. During the war, uranium production had been exclusively a government production at MED facilities. Realizing that a domestic program for uranium production would be capital to build a stockpile of atomic bombs, the AEC made it a top priority to encourage private businesses to prospect and mine the ore. The Act designed a system that seemed to be a win-win situation because it guaranteed private producers a market while maintaining the government's exclusive supply chain. In 1949, the AEC issued a booklet titled *Prospecting for Uranium* to describe the different types of ore deposits, the major uranium-bearing minerals, selling procedures, laws, regulations on mining claims, and instructions on how to use a Geiger counter. This booklet was the first in a long series of publications as magazines in the 1950s took over and presented uranium prospecting as a hobby, with headlines such as "From Rags to Riches with Uranium," "Build Your Own Geiger Counter," or "Uranium for the Amateur and Hobbyist."[9]

The AEC's demand was so high at first that contracts were not signed with the prospectors who found the deposits but with big companies such as the Vanadium Corporation of America, the Anaconda Copper Company, the Phillips Petroleum Company, and the Kerr-McGee Oil Industries, who bought out small prospectors. Not only did the AEC maintain a monopsony on purchasing uranium ore, but it also held a monopoly on providing enrichment services. Private companies that were licensed to build power plants had to lease their fuel from the AEC monopoly.

Grants came to life at the rhythm of the uranium industry. Funds for city projects, such as building new schools, came from two main sources of income: property taxes from the uranium companies and federal loans. A local branch of the New Mexico State University was established in 1968 in cooperation with the Grants Municipal Schools and offered a program in underground uranium mining training. Thus, "Grants voluntarily linked itself to both the uranium industry and the Federal Government in a type of corporate and governmental colonialism."[10]

The transition out of the monopsony toward a system of allocation, protectionism, and subsidies between 1962 and 1970 posed a problem because private buyers could not provide enough demand to compensate for the AEC's withdrawal from the market it had created. As supply kept increasing and demand stagnated, ore prices and sales were no longer guaranteed. The stockpile of uranium for atomic weapons was sufficient; weapons research refocused on hydrogen and later on missiles. At the same time, a growing pressure to ban atmospheric tests also impacted the needs for uranium. Therefore, the AEC's purpose was no longer to accelerate the industry's development but to save it from collapse "by making it easier for private companies to own and operate nuclear power materials and protecting U.S. producers from foreign competition." The Private Ownership of Special Nuclear Materials Act of 1964 made it legal for uranium companies to sell yellowcake directly to nuclear power plant customers by permitting private ownership of fissionable materials and it also introduced protectionist measures.[11] Energy companies, having replaced small prospectors, began drilling deeper to find more ore, but the demand of power plants was still not enough to absorb production. During this first bust phase, Grants experienced a wave of layoffs, high unemployment, and drop in property value and income.

Ten years after the start of the first bust phase, uranium prices increased again and "peak production was attained in 1978, with a record yearly production of 9,371 tons of uranium, that was shipped to mills and buying stations."[12] The oil crisis of the 1970s stimulated demands for alternative energy, and many power plants had reached the end of the long preparation process and were ready to begin operations. On March 18, 1975, a report of the Committee on Nuclear Energy for the state of New Mexico read, "New Mexico currently supplies more than 40% of the uranium necessary for the U.S. nuclear power industry. . . . Because New Mexico will be called upon to continue to supply research, development, and materials for this valuable energy resource, the state should *reevaluate* its position relative to the needs of the nation."[13] The word "reevaluate" refers precisely to the state's role throughout the Cold War as it regularly reevaluated its activities according to the government's and the nation's needs. Grants' population kept growing, and a new county was created for organizational purposes out of the western part of Valencia County in 1981. The name of Cibola County, chosen in reference to Coronado's Seven Cities of Cibola, was a nod in the direction of the bounteous uranium industry.

Due to growing demand, the Joint Committee on Atomic Energy then considered stopping the ban on foreign uranium importations. They "provided for the gradual elimination by the end of 1983 of restrictions on its enrichment of foreign-source uranium for domestic use."[14] The industry suffered from the gradual opening to foreign producers and growing environmental and safety concerns that began to have an impact on nuclear energy activities. In the wake of the Three Mile Island accident of March 1979, the Nuclear Regulatory Commission (NRC) required energy companies to put plans for new power plants on hold. The companies who abandoned their plans put their yellowcake back on the market at bargain prices.[15] In the bust of the 1980s, unemployment in Grants reached 30 percent. The population dropped from twenty thousand to ten thousand when it had been projected to reach one hundred thousand. The people who left town sold their houses for a fraction of their cost and many of the town's businesses closed.

In 1984, Secretary of Energy Donald Hodel, to whom Congress had assigned a comprehensive review of the domestic uranium mining and milling industry, concluded that the industry was no longer viable. Domestic producers could not survive the collapse of the market without the government's protectionism. Kerr-McGee at the head of the Quivira Mining Company—Quivira, like Cibola, refers to the Spanish conquest and the myths of opulent Native communities—was the last major uranium employer to close its mines and mill at Ambrosia Lake on January 16, 1985. Two years later, Western Nuclear, a uranium mill tailing site in Jeffrey City, Wyoming, filed suit against the DOE in federal court "alleging that DOE's failure to impose restrictions on the enrichment of foreign uranium for use in domestic facilities constituted a violation of" the Atomic Energy Act of 1954. The court ruled in favor of the DOE, which had argued that "the imposition of restrictions on DOE's enrichment of foreign uranium would not assure viability" of the industry.[16] Jeffrey City became a ghost town. Grants, however, managed to recover somewhat from its boom-and-bust uranium history thanks in part to tourism, a uranium mining museum, and the construction of three state prisons. Nevertheless, the area still looks much depressed.[17]

Since the substantial increase in uranium prices between 2004 and 2008, uranium mining companies have shown renewed interest in New Mexico's deposits in the Grants Mineral Belt. A report prepared for the New Mexico Environmental Law Center in 2008 reads: "Now uranium mining companies and other business interests are promoting renewed uranium mining as a potential source of thirty billion dollars and almost 250,000 jobs." The report states that these claims are a "gross exaggeration," that "New Mexico knows from experience with copper and uranium that metal mining is economically unstable," and that "important environmental and social costs must be considered when evaluating the commercial economic benefits of renewed uranium mining."[18] The report shows that the debate on the industry's revival in the state is ongoing and that, despite its economic development in the past century, New Mexico is still facing the same economic dilemmas. However, a lesson has clearly been learned from the story of Grants and the risks involved in a region's relying completely on a single extractive industry, no matter the amount of government help and capital invested.

During a meeting of the Southwest Indigenous Uranium Forum in September 1993 in the village of Paguate, overlooking the Jackpile mine, Acoma Pueblo activist Manuel Piño talked about the dangers of economic promises. The mining companies had argued that the miners were trained in a marketable skill, but these skills were worthless once the mines had shut down. "Economic incentives did not necessarily improve our way of life," he said. As land was lost and wages were earned to buy the products of modern America, "social problems" such as alcohol and drug abuse, domestic violence, higher school dropout, and suicide rates appeared in the Pueblo.[19] In this case, federal government sponsoring contributed to the creation of a neocolonial exploitation of resources and people becoming dependent, conquered, and still suffering from the consequences of exploitation.

The Challenge of Diversification: Peaceful Ventures

As illustrated by uranium production, the exceptional circumstances of World War II and later those of the Cold War were a sine qua non for the development of New Mexico's federally sponsored economy. Funds were determined by national security needs, and those needs were contingent on the circumstances created by war. While the country had never maintained a large standing military establishment in peacetime, the Cold War was different in that R&D installations could not be demobilized at the end of the conflict. As a traditional recipient of Army facilities such as forts, bases, and Navy shipyards, the West had a history of economic relations with defense industries, but this time, the nature of the war prefigured more permanent benefits for the areas around the facilities.

Bombs and test series in the nuclear era ensured work from one end of New Mexico's nuclear chain to the other: Grants mined and processed uranium; LANL designed new weapons and participated in test preparations; SNL produced and assembled the weapons; and White Sands pursued rocket and missile R&D. As a result, all these installations and the myriad of small firms involved in the defense effort as contractors, thrived thanks to the nation's defense budget, but they were also tributaries of events happening far beyond the borders of New Mexico. Money influxes and contract numbers fluctuated with changes in global politics. The budgets of the Laboratories, for instance, had to be revised several times. In Sandia, the budget for fiscal year 1951 was submitted in May 1950 based on assumptions current at that time but had to be revised upward because of the outbreak of hostilities in Korea.[20] In Los Alamos, the Zia Company's financial situation depended entirely on the AEC, which reimbursed all the company's bills and salaries to its employees. Calculation of "the Zia budget" was based on an estimate of the cost of operating the town and the technical areas. By the mid-1950s, Los Alamos represented a $250 million federal investment.

After a first booming stage in the arms race in the 1950s when programs and contracts poured in to prepare the Greenhouse and Tea-Pot test series and to produce the hydrogen bomb and the first missiles, the state's overdependence on military money revealed itself as research was redirected to ground war in Vietnam. The test moratorium between 1958 and 1961 exposed the possibility that the weapons laboratory might have to face serious budget cuts and could possibly close down. Concerns arose in the United States about the economic impact that disarmament might have on regions that heavily relied on defense industries and federal funding.

In December 1965, the US Arms Control and Disarmament Agency commissioned economic consultants with a regional study to "enhance the ability of the Federal Government, in collaboration with state and local governments and with industry and labor, to deal with the economic consequences of arms control and disarmament." In their overview of the state's economy, the authors write:

> By far the largest basic industry in the state is federal activity, especially that portion which represents defense expenditures. Nearly one-half of all employment and income derived from basic activity in the state is estimated to be due either to direct employment of civilian and military personnel in defense establishments or Federal Government purchases of goods and services related to the defense effort.

One of the issues raised by the study was a lack of alternatives to the defense industry, for which the authors blame the discrepancy between this "highly sophisticated" industry and New Mexico's "traditionally underdeveloped economy" that prevented "substantial expansion of existing or development of additional non-defense basic industries." As a result, the whole state ran the same risk as in the Grants area of relying heavily on one economic sector without a so-called plan B.

The defense sector—DOD, AEC, and the National Aeronautics and Space Administration (NASA)—represented fifty-seven thousand employees and 20 percent of total employment in the state. A decrease in defense spending would result in layoffs, and the affected people—military and civilian personnel—would most likely desert the state as they were transferred or assigned to a job elsewhere so "not even savings and unemployment compensation benefits would be entering the local economy."[21] The revenues of the state government would not be overly affected, but those of the municipal and county government dependent on the local economy would. In 1967, the dangers of excessive reliance on defense jobs were felt with the closing of the Walker AFB that employed 3,200 military and 350 civilian workers. A year later, the AEC also put an end to its cooperation with ACF Industries, a nuclear weapons manufacturing facility in Albuquerque that employed about 2,700 people in 1964. The two combined represented the significant loss of 2.4 percent of the state's nonagricultural jobs.[22]

The study recommended economic adjustment by developing alternative basic export activity, stimulating the private sector, expanding existing federal nondefense, and converting existing facilities to nonweapon purposes. However, "a lack of concerted effort to meet possible adjustment problems indicates that the effects of reduced defense expenditures . . . could be felt in full measure unless there were changes in attitudes and increased planning." This conclusion was reached after a series of interviews with leading state and local officials, business executives, labor leaders, economic development directors, university officials, and military commanders. The authors note, "Absentee corporate officials often felt that their New Mexico operations were not central to their success." In other terms, large corporations were prepared to desert the state altogether in the event of a major decrease in government contracts. Evidently, nuclear science had revolutionized the state within two decades, but the new industries could vanish just as quickly, leaving the region to fall back into its initial state of underdevelopment. Last, the study underscores the distance that separates New Mexico decision makers from the center of federal policymaking, leaving them "unaware of the various forces affecting national security and arms control policies." This observation confirms that a direct connection was established between decisions made in Washington and the fate of New Mexico's industry and, by a ripple effect, that of its residents. Hence, the last recommendation was that public understanding of the interplay between reductions in defense spending and the US policy should be increased.[23]

By 1960, New Mexican officials had come to realize the dangers of dependence on the local nuclear weapons complex. Of the AEC's $900 million budget, one-sixth was spent in New Mexico where the agency had three offices—two in Albuquerque and one in Los Alamos. The AEC directly employed more than six hundred people in these offices, and its contractor employment in both towns was about fifteen thousand, with an annual total payroll of approximately $152 million. Senator and Chairman of the Joint Committee on Atomic Energy Clinton P. Anderson commented, "The laboratories we have built for weapons can, and must be shifted to peaceful work once the great demand for weapons is over" because "to permit this investment [facilities, plant, organization, and highly skilled and highly dedicated scientific, engineering, and administrative manpower] to be dispersed and not utilized would be a waste of magnitude that we could not, in the national interest, afford." Anderson believed that the Sandia Corporation could expand by redirecting research toward "the peaceful atom and the conquest of space." Moreover, efforts "of forward-looking citizens" were being made to guarantee that "any adjustment of military spending" would "evolve without affecting the overall economy" so that "Albuquerque will be prepared to carry on without the enormous payroll, and switch over to a completely civilian economy."[24]

Yet, the buildup continued for many more years. The work on missiles continued after the escalation of tensions in the wake of the Cuban Missile Crisis of 1962. Excavation projects and underground tests replaced the atmospheric series after the Test Ban Treaty of 1963. The "space race" culminated with Apollo 11 landing on the moon in 1969. The US arsenal would eventually peak in 1987 with more than thirteen thousand nuclear warheads. New Mexico's defense-related facilities and research centers were able to carry on with their operations. Nonetheless, as part of attempts to diversify the state's economy, efforts were made to redirect research on nuclear devices toward peaceful utilization.

New Mexico was chosen for two experiments of the Plowshare program, which was the application of nuclear explosions to peaceful purposes inspired by the ideals that President Eisenhower had expressed in his Atoms for Peace speech in 1953.[25] The experiments generated the anticipation of local energy companies but both were utter fiascos. One objective was to show the "friendly" side of the atom to the public, but the Gnome and Gasbuggy tests in 1961 and 1967 proved that diversification would be a challenge and that the public would be called upon to bear the risks of the tests for private profit. In both cases, there was almost no opposition except a telegram sent by a chemist from Los Angeles, Dr. E. H. Bronner, who "charged that the [Gnome] nuclear blast would destroy Carlsbad's mineral deposits."[26] The experiment was presented as a gamble: if it was a success, the commercial gains would be worth the risk of destroying the area's first industry.

In 1961, many locals in Carlsbad were eager to see nuclear science introduced in the vicinity if it meant it would bring a decisive economic boost. The plan was to use the explosion to create an underground heat reservoir that would vaporize in the salt bed, and the steam would be transformed into a power source. The area was chosen for its technical and geological features, as well as for low population density and the fact that the land was under government control. After President John Kennedy gave the green light to the project, an article reported local reactions such as that of Roger Jenkins who said, "I'm tickled to death. I think it will help our situation." He added, "He felt the detonation and subsequent activity would boost the economy of the town even more than it has been helped by pre-Gnome activity." The article also quoted Ed Skinner, resident manager of the Potash Division of the International Minerals and Chemical Corporation, who fervently declared, "It will give the Russians something to think about."[27] From the time Gnome was announced in August 1958 until the detonation three years later, excitement built up. The economic prospects of a new way to produce electricity seemed phenomenal. In 1959 Mike Hill wrote in the *Carlsbad Current-Argus*, "The nuclear bomb, a symbol of destruction, may prove to be a boon for the Carlsbad area in the near future." The project "could be responsible for the location of an electrical power production plant" and be "used as a production center for producing and recovering valuable isotopes."[28]

After seeing the prosperity and stability federal nuclear programs had brought to Los Alamos, Alamogordo, Grants, and Albuquerque, local residents were hopeful that their own experiment would bear similar fruit. On the other hand, farmers were afraid of fallout for their fields, miners feared for the potash mines, and tourism professionals feared for their famous caverns. The panel of expert scientists commissioned by the AEC to study these various issues concluded that the explosion would be fully contained in the salt beds, but the agency still cautiously "signed contracts with the seven mining companies in the Carlsbad potash basin to provide reimbursement for loss in production as result of mines being shut down" during the project and "provide for payment . . . for services supplied by the potash firms during the pre- and post-shot mine surveys."[29]

Three hundred visitors, including officials and newsmen from ten nations, came to witness the test on December 10. The blast produced a cavity about 75 feet high and 150 feet in diameter. US Representative Thomas G. Morris of New Mexico addressed the audience after the detonation to emphasize the historic significance of the moment: "As Alamogordo became the symbol of the beginnings of the A-bomb, so will Carlsbad symbolize to all mankind the beginning of peaceful uses of nuclear explosives."[30] New Mexican politicians thus proudly sought to further the state's connection with nuclear history. One unexpected and worrisome consequence of the test, however, was the radioactive steam that emerged from the elevator shaft. But experts again assured there was no danger.

Gnome was deemed a success even though its main aim—to produce electricity—was not attained. In the end, the most valuable information that resulted from the experiment was the geological and seismological data, which "contributed significantly to understanding the geological structure of the United States and probably will result in a reinterpretation of the properties of the earth's crust and mantle east of the Rocky Mountains."[31] Notwithstanding its geological value, the second test that had been planned at the same location was canceled.

Six years later, Gasbuggy was presented as the first joint federal government–private industry experiment. El Paso Natural Gas was the AEC's industrial partner for the project that was meant to stimulate natural gas production from low-permeability formations. It was anticipated that "the explosion [would] vaporize the rock around it and create a subterranean chamber deep below the surface. Then the roof of the chamber [was] expected to fall in, creating a 'chimney' or area of broken rock."[32] The blast was to take place in the San Juan Basin, on the property of El Paso Natural Gas, fifty-five miles east of Farmington and just west of the Jicarilla Apache Reservation. The nation approved the test because most of their income was derived from oil and gas, so a road was built through the reservation to the project site.

This test's conclusion was not brilliant either. The gas flow was only slightly increased and had become radioactive. Fallout concerns at the NTS and revelations about the consequences of strontium-90 absorption (such as bone cancer and leukemia) had begun to raise fears among the public by that time. They were not willing to take the risk of using radioactive gas, no matter how low the level of radioactivity.[33] Both blasts were disappointments and greatly impacted the environment as the release of radionuclides contaminated both air and soil to varying degrees despite the cleanups. Both tests illustrate how the government and big corporations negotiated with locals, such as the Carlsbad mining companies or the Jicarilla Apache Nation, to obtain their support. In both cases, expectations ran high because of the struggles that local communities had been facing and disappointment was, accordingly, great.

Attitudes Toward the Nuclear Industry

Federal spending in New Mexico doubled again in the 1970s with the arrival of new programs such as the space shuttle and laser weaponry, and continued research on nuclear energy sources.[34] With each program, jobs were created, and more people came into the state. The economic map of New Mexico was redrawn, presenting substantial disparities from one area to the next and an inverse correlation between federal spending and welfare. Consequently, local elites, city boosters, businessmen, and politicians strived to attract more and more funds to ensure economic growth and employment. Various interest groups learned how to use global politics and events such as Sputnik or the missile gap to pressure the administration in power for more funds in their particular field whether it was business, education, or science.[35]

Albuquerque boosters used the introduction of the Sandia Laboratory to advertise the Duke City as a Mecca for technology and a diverse metropolis, welcoming industry and growth. The new population of white-collar, upper-and-middle-class families in the so-called science suburbs of the Northeast Heights depicted Albuquerque as a dynamic, modern city characterized by scientific research, higher education, and a strong federal presence. Historian Robert T. Wood explains that "growth of government activity had led to overlapping of city, county, state and federal functions, and the several layers of government had become progressively more intertwined."[36] The government even intervened in real estate by granting federal mortgages to families who arrived to work at Sandia. By 1980, 40 percent of the population of Albuquerque had resided there fewer than five years and four inhabitants in five had been born elsewhere.[37] In order to cope with the growth, city leaders had to ask for the help of city planners, acquire more land, and improve the roads. According to Marc Simmons, "Albuquerqueans were content to let technocrats and urban planners chart the destiny of their city." Simmons describes these individuals as people who "think almost wholly in economic terms, equating unbridled expansion with automatic prosperity and seldom paying more than lip service to such humanistic considerations as municipal beautification and historic preservation."[38] In the 1960s, realizing their vulnerability in the face of federal cutbacks, businessmen and city-boosters reacted by organizing groups such as the Albuquerque Industrial Development Service to broaden the local industry base.[39]

In parallel to these boosters' efforts, New Mexican politicians from the late 1940s to the present have supported the industry, sometimes in the face of controversy. Western politicians in Congress had traditionally exchanged their votes on national controversies for support on issues that mattered most to them, such as land and resource policies. Getting gradually accustomed to federal help to keep their states on the sharp edge of science and technology, they pushed for more nuclear programs despite the risks involved in having their local economies subject to outside forces. One example was Congressman Dennis Chávez, who wielded his influence throughout his career to attract public investment. Chávez became senator in 1935 and first used his power to bring relief programs such as the CCC and National Youth Administration to New Mexico. Maria E. Montoya attributes to him the arrival of additional postwar federal facilities—WSMR, SNL, and Kirtland AFB—and the increase in funding. In 1960, when he served as chair of the Senate Defense Appropriations Committee, he recommended a budget $1 billion higher than the already exceptionally high budget recommended by President Eisenhower. At the time he also worked with Representatives Joseph Montoya and Tom Morris to help Albuquerque and New Mexico "continue their roles in the space age."[40] Montoya writes that Chávez thus showed "pragmatism since he sought to funnel much of that spending to his own state's constituents" and adds that "thousands of New Mexicans went to work for the Federal Government in a myriad of jobs brought to the state by Chavez's deft political skill in the Senate."

New Mexico's specific role within the national nuclear weapons complex made the ability of its representatives to tip the scales in the right direction on important issues pivotal to maintain high employment. Congressmen had a double-edged political game to play, for if New Mexico's role in the nuclear industry could be used to increase their influence in Washington, Congress could also threaten to close the funding faucet. It is interesting that through her analysis of Chávez's political career, Montoya addresses colonialism. She notes, "During the nineteenth century, New Mexico maintained a colonial relationship with the rest of the United States because of its dependence on eastern and foreign investment. Still, New Mexico in many ways remained in that same colonial stance throughout the late twentieth century."[41] The main difference between the two periods was that its main source of investment was now contingent on defense budgets, and politicians had to exert pressure on public institutions for jobs.

Senator Clinton P. Anderson, quoted earlier, was another New Mexico elected proponent of the nuclear industry. As chairman of the Joint Committee on Atomic Energy from 1955 to 1961, Anderson aggressively promoted military nuclear development, including increased production of fission bombs and fissionable material, development of the Super, and opening of more uranium mines. He passed the 1957 Price-Anderson Nuclear Industries Indemnity Act, which limited the liability of the nuclear industry in the event of a nuclear incident. He was a strong supporter of the Space Program as well.

Promoting the nuclear industry became more delicate in later decades as the public grew suspicious of the impact nuclear jobs and employers could have on health and the environment. Yet, even in the face of mounting discontent, local politicians defended the state's economic backbone. Senator Pete Domenici, who was elected in 1972 and served until 2008, actively engaged in the promotion of the Labs with increased funding and programs to improve the testing of nuclear weapons without use of physical underground testing. He declares in his book: "My ultimate goal is that in the year 2045, one hundred years after the detonation of the first atomic bomb and the birth of the nuclear age, the world will evaluate the role played by nuclear technologies and conclude that their overall impact was strongly positive."[42]

Meanwhile, the state government was also called upon to make the most of the nuclear industry's opportunities. A study of the Committee of Nuclear Energy for the Governor was undertaken in March 1975 "to provide state policy organization with the information needed for formulating a rational government position" and to encourage the state to expand its activities into the uranium fuel cycle for nuclear power plants in the United States and abroad. The committee concluded that "the economic benefits to the state . . . are so great that they must be exploited." Among the benefits, the opportunity to "train and provide employment for its citizens in an expanded nuclear industry" was underlined. The study's major recommendations were to "launch a vigorous program to support increased uranium exploration and production, both in recognition of national need and for economic benefit to the state" and to "aggressively encourage either private industry or the Federal Government to construct a centrifuge uranium enrichment facility in New Mexico in the early 1980's."[43]

In January 1977, the Bureau of Business and Economic Research of the University of New Mexico conducted a survey by telephone with almost 2,800 people "to determine the attitudes of New Mexicans toward expanding different parts of the nuclear power industry in the state." The results were thus introduced: "The main finding of the survey was that New Mexicans favor establishing more of the nuclear fuel industry in the state." However, taking into account the percentages of the last four possible answers ("don't care, disapprove, strongly disapprove, and insufficient information"), the conclusion is not so unequivocal for questions directly concerning New Mexico: 43.3 percent disapproved or did not have an opinion about locating more of the nuclear fuel industry in the state; 74.8 percent disapproved or did not have an opinion about storing waste materials from the nuclear power industry; 46.8 percent disapproved or did not have an opinion about installing uranium conversion, enrichment, and fabrication plants; and 51.5 percent were unsure or believed the benefits to be gained from the nuclear fuel industry in New Mexico were not greater than the possible safety and environmental problems.

Such figures demonstrate the wariness, and perhaps also the anxiety, of New Mexicans toward expanding the nuclear industry in their backyard, notwithstanding the economic returns. Another finding of the survey was that "the proportion of persons saying they had insufficient information to answer questions declined almost uniformly with education level." Furthermore, there was "a tendency for persons with higher educational attainment to be somewhat more opposed to additional nuclear industry installations in New Mexico."[44] In plainer terms, considering that many of the state's highly educated people were scientists newly arrived from other states to work at the Laboratories and that native New Mexicans were more likely to fall under the category of the less highly educated, locals were generally less informed and approved more readily the expansion of the nuclear industry. This touches on the fundamental link between knowledge and the acceptance of risks. Populations who are kept in ignorance of the possible consequences and are desperate for employment opportunities will be more prone to accepting an economic and environmental gamble.

Ten years after this study, the state still relied heavily on federal funds and jobs despite attempts at diversifying the economy. The number of federal employees in New Mexico in 1987 totaled 79,338 (calculated by adding federal civilian employees, military service men and women stationed at defense installations, SNL and LANL employees, other DOE contract employees, and all other defense contractor employees who worked on defense installations around the state). Combined with other state and local government employees, "one concludes that almost 47 percent of all employment in New Mexico can be attributed to governmental activity." The other activities were remnants of the state's preindustrial past: tourism (8.5 percent of the state's nonagricultural jobs), manufacturing (7.3 percent), mining (2.8 percent), and agriculture (3.6 percent of all jobs).[45]

Now, in the twenty-first century, the state's excessive dependence on federal money remains an issue that continues to contribute to its vulnerability. Geopolitics on an international scale can still impact the local economy, but domestic matters can be problematic too. In 2013, a two-week governmental shutdown, which lasted from October 1 through October 16, exposed this fact glaringly when LANL and SNL were on the verge of closing their doors. New Mexico was among the states hit hardest by the crisis. The (partial) shutdown occurred because of the controversial Patient Protection and Affordable Care Act (PPACA), known as Obamacare. Right-wing representatives in the Republican-led House of Representatives "attached a provision to a spending bill that required eliminating funding for the implementation of the PPACA in order to fund the rest of the U.S. Federal Government."[46] The bill was sent to the Democratic-led Senate, which stripped out the provision before sending it back to the House. This created a funding impasse and Congress did not pass a law to appropriate funds past September 30, 2013.

Kirtland and Holloman AFB placed 1,074 and 422 civilian employees on furlough, respectively. The Labs, which are run by independent contractors but with public funds from DOE and the National Nuclear Safety Administration (NNSA), had "carry-over funding left from previous years' budgets to keep working for at least a short period of time." Nonetheless, both LANL and SNL announced their shutdowns for October 18 and 21. Their closing down concerned close to twenty thousand employees. The Federal Bureau of Land Management freeze also impacted the oil and gas industry "due to the large amount of federally owned land"—roughly 35 percent of the state's total acreage. At New Mexico's eleven national parks and monuments, workers had to be furloughed as well. In 2012, roughly 24 percent of the state's nonagricultural workers were employed by federal, state, or local governments.[47] The shutdown further highlighted New Mexico's dependence on government jobs. This fact raised again the awareness of local journalists who pointed out that more than one-third of the state's gross domestic product came from federal spending. Some have addressed the question of how to make New Mexico less vulnerable to federal government shutdowns.[48] Their answers consistently focused on diversifying New Mexico's economy, supporting local entrepreneurship, and improving performance in education. These current challenges clearly reflect the recurrence of the state's disquieting situation.

On a final note, such federal involvement is all the more noteworthy or surprising in the context of the United States. It is generally recognized by political scientists and historians that US politics are shaped under the pressure of group and individual interests and that the executive branch has limited capacity for independent action in domestic matters. Fear of too strong a government is reflected in the checks and balances principle at the core of the Constitution. Therefore, there have been many instances in American history when the federal government was reminded it did not have authority on domestic matters. In 1935 and 1936, Roosevelt's National Recovery and Agricultural Adjustment Acts were declared unconstitutional by the US Supreme Court in *Schechter Poultry Corp. v. United States* in 1935 and *United States v. Butler* in 1936 because, even though the economic crisis was a national emergency, it did not justify enlarging the executive's constitutional authority as contemplated in the New Deal. World War II presented a different context, which carried over into the Cold War to some extent. Some issues of national security were made exceptions to the rule of a weak executive. When confrontation with the Soviet Union emerged as a national security threat, the authority of the traditionally weak executive in peacetime grew, using a key ingredient of wartime operations: secrecy.

Eight

Cloaked in Secrecy

The complicated interactions among corporate interests, government funding, and a pervasive use of secrecy cemented New Mexico's nuclear weapons complex from the onset. The same intricate relations remain to this day. LANL, for instance, is funded by the Department of Energy (DOE) and National Nuclear Security Administration (NNSA); it is managed by the University of California; and it has relationships with each branch of the US military, as well as corporations, industrial suppliers, and subcontractors. The Cold War fomented a culture of secrets that was at the heart of these interactions and enabled the concealment of some of the most damaging impacts of the Manhattan Project. Official, selective discourses modeled the public's zeal to believe in the miraculous prospects of atomic science and in the necessity of the escalation in armament by putting forward a binary, simplistic worldview and keeping silent on the darker sides of the military buildup. This culture stimulated the fervor and commitment of all participating individuals at the Labs, research centers, and testing sites either by clouding the downside of their work or by infusing them with the new philosophy of nuclearism—that is, the faith in nuclear weapons to maintain national security.

In Los Alamos, secrecy was pushed to its paroxysm from the start, but after the atomic bombings, it was also the place where the first antinuclear movement developed. The pursuance of secrecy was the primary point of contention against which the first activists argued. The movement was led by scientists of the Manhattan Project who believed that secrets would be responsible for the escalation of arsenals. In his speech to the Association of Los Alamos Scientists (ALAS) on November 2, 1945, Robert Oppenheimer declared, "Secrecy strikes at the very root of what science is and what it is for."[1] To defend their ideas, the Los Alamos scientists began giving public speeches, first in the vicinity, in Santa Fe and Taos, explaining the principles of atomic fission and insisting on the importance of international control of atomic weapons to avoid an arms race.

The movement was successful in getting the public's attention at first. They used their influence to infuse a fear of the bomb in the public's mind, hoping it would be a catalyst for a move out of secrecy and compartmentalization toward international cooperation. But the strategy backfired. Instead of prompting rational cooperation between nuclear powers, fear produced hatred, blind terror, and a thirst for American superiority. The scientists' movement died out at the turn of the 1940s and 1950s, its ideas obliterated by the ideology of massive retaliation. The story that best personifies the rise and fall of the movement is that of Oppenheimer. His views (including his opposition to the Super program) and former communist connections at the height of McCarthyism cost him his clearance after a security hearing in 1954, and he disappeared from politics. His political demise was covered nationally in the news. In New Mexico, the message to the Laboratory's employees seemed to be, "If you disagree with official policy, your work at the Lab could be terminated and your career ruined."[2] Following the first upsurge against the Atomic Energy Commission's (AEC) policies, it took until the 1960s for antinuclear activism to be rekindled in the United States. Protest again gained momentum with the rise of the environmental movement.

Hush Hush, Bombs are Being Born

Despite the scientists' fight, Los Alamos became the epitome of atomic secrets. To this day, the community built in secrecy has carefully maintained an aura of mystery for the pleasure of tourists and visitors who relish the anecdotes of wartime secrecy and compartmentalization. Perhaps one of the most amusing aspects of secrecy was the coded lexicon inside the lab: atoms were "tops," bombs were "boats," an atomic bomb was a "topic boat," plutonium was "product," and uranium fission was "urchin fashion." Proper names followed the same rule: Oppenheimer was James Oberhelm, Enrico Fermi was Eugene Farmer, and Niels Bohr was Nicholas Baker.[3] Secrecy made the little community sensational. The drastic restrictions on communications and the strict rules governing the community were the main focus of the first articles on the secret town. George Fitzpatrick, reporter for the *New Mexican Magazine*, wrote in September 1945, "Even though many Santa Feans and many relatives of Santa Feans worked there, the oath of secrecy was kept, and New Mexicans generally accepted the fact that 'Los Alamos' was a taboo subject. . . . the place just did not exist, even though convoys of trucks made frequent trips through Santa Fe and government buses brought workers into Santa Fe daily for brief leaves or shopping trips."[4] Famed scientists were rarely recognized with the exception of Einstein who was rumored to have been spotted there—even though, incidentally, he was never invited to participate in the Project because the FBI feared his socialist connections.

The rumors circulating in Santa Fe about the Project during the war were, in fact, as extravagant as the Project was colossal. They included research concerning gas warfare, rockets, jet propulsion, spaceships or "death rays" of an indeterminate type, windshield wipers for submarines, a camp for pregnant WACs, and even a "Republican internment camp" during the 1944 presidential election campaign.[5] Interviewee Rubén Montoya recalls the hearsay that circulated about the Hill: "They said they were making submarines and they would float them down the Rio Grande."[6] Because he was afraid that one of these murmurings might get closer to the truth, Oppenheimer asked scientific librarian Charlotte Serber to go to Santa Fe and spread the rumor that the Lab was building an electric rocket.[7] People wondered whether the gates were meant to keep people inside or keep them out. The Manhattan Engineer District's (MED) Counter Intelligence Corps (CIC), which was in charge of keeping the project classified, encountered difficulties, such as a Cleveland reporter who, after spending a vacation near Los Alamos, had hurried home to write an exciting story titled "The Forbidden City." He intended to write, "The Mr. Big of the city is a college professor, J. Robert Oppenheimer, called the 'Second Einstein' . . . widespread belief is that he is developing ordnance and explosives . . . others . . . will tell you tremendous explosions have been heard."[8] The CIC hurried to hush him.

Because of how little information was available, simply sending a reporter through the gates to depict life beyond the fence was handled as breaking news. An "open doors" event was organized for the tenth anniversary of Trinity. In July 1955, LASL orchestrated a tour of the technical areas, which had been up to then rigidly restricted to cleared personnel, for reporters, staff, and employees' families to discover the nucleus of the laboratories, view top-secret experiments, and attend a science fair. General Leslie Groves organized tours of the Trinity site with Trinitite as souvenirs for visitors. In order to reassure reporters that radioactivity in the crater was safe, Groves ordered Patrick Stout, an Army counterintelligence agent, to descend into it to be photographed. Although Stout seemed to come out unscathed, he was probably exposed to a dose of about a thousand roentgens. He died of leukemia in 1967.[9]

One less well-known aspect of wartime secrecy is that it was another indicator of the way the indigenous population was perceived. In view of the precautions that were taken to make sure that Santa Feans would not make guesses too close to the truth of what the District was doing on the Hill, Project leadership was surprisingly unconcerned with the local workers who spent most of their days amid atomic physicists. Phyllis Fisher wondered at this differential treatment; she writes,

> Curious, wasn't it, that our project had to be kept secret from the Caucasian population of Santa Fe and yet the Indians and Spanish-Americans were bused daily to the project? A secret community was being infiltrated daily, and no one was concerned. Was it because these people could be trusted more than the white population? I don't think so. Was it because they were considered too unimportant to be a security threat? Was it because we hadn't thought about them as people, as individuals? We needed their service to take our coal, clean our houses, collect our garbage, and twist wires in the tech area. Wasn't it ironic that an Indian chieftain was ladling out soup and serving steak in our cafeteria?[10]

Fisher's comments underscore the invisibility of the indigenous peoples and the absence of individuality when referring to local participants throughout the process of scientific conquest. Their presence mattered in terms of cheap, close-by, and discreet manpower but was not envisioned as a potential threat, such as a security leak or objection to the Lab's activities, because they had neither the knowledge nor the agency. Furthermore, as argued earlier, these populations had everything to gain from the arrival of the new industry, and the mechanisms of nuclear neocolonialism appear in Fisher's thoughts: conquerors have traditionally viewed the conquered as a blind, uneducated group that needed to be guided and enlightened, and that could not acquire the same skills and knowledge autonomously (cf. the three-people murals).

Constraints of secrecy and security aside, the 1950s population of Los Alamos was content with living in a gated town that protected the peculiar composition of an elite community. The system of segregated housing, which guaranteed that the best residential areas were assigned to scientists and engineers with the highest degrees, concentrated the elite toward the center and kept the non-Anglo working class on the outskirts. This strat-ification drew an urban landscape based on education and ethnic discrim-ination. The fences separating the town from the poorer region outside further emphasized this pattern. A poll in September 1954 showed that 81 percent wanted to keep the outer fences closed.[11] "Safety" and "privacy" were the two reasons invoked by residents for keeping the gates and guards in a survey in 1955. Urban studies and planning specialist Carl Abbott analyzed these terms. He writes, "Since the only invaders in the mid-1950s were likely to be handfuls of tourists and local New Mexicans, it is hard not to read these terms as code words for class and ethnic prejudice—attitudes as deeply embedded in the country's sci-tech elite as among any other Americans of the 1950s."[12] Nonetheless, the AEC opened the gates and removed the guards on February 18, 1957.

The polls prior to the opening of the gates revealed one of the town's paradoxes: the Laboratory worked on the most destructive weapons on earth and dumped toxic waste in the surrounding canyons, yet residents still considered Los Alamos a safe place. One scientist said to a reporter in a 1948, "[Los Alamos]'s been so good for the children. They seem much stronger, and they're not as high-strung as they were in the city. Those mountains are soothing. They're permanent. Nothing could ever disturb those mountains—except maybe our bombs."[13] The paradox between a "soothing" environment for children and the Lab's dangerous work largely stemmed from ignorance and undisclosed information about the consequences of nuclear weapons work. After all, as Gerard DeGroot ironically points out, "Baby boomers grew up believing that a wooden desk would protect them from a 10-megaton blast."[14]

However, incidents happened and some of them were serious enough to produce anguish among Los Alamosans. The first casualty was Lewis Slotin, who was chasing the dragon—an experiment to measure criticality—and separated hemispheres of plutonium with his bare hands on May 21, 1946. He died nine days later and was buried in a lead-lined casket. The second victim was Harry Daghlian, on September 15, 1946, who died of acute radiation syndrome. The newspapers only mentioned severe burns due to an industrial accident. Cecil Kelly, a LASL plutonium technician, thirty-eight years old, was the third victim in December 1958. As he was stirring a vat of radioactive waste with an automatic paddle, he absorbed 12,000 rems and died on the following day.[15]

Less tragic incidents occurred, which clashed nonetheless with the image of a child-friendly environment. Children would incorporate words such as "contaminated" into their games or staked their building blocks into high piles and then dashed into them screaming "I'm an atomic bomb!"[16] They would play with unexploded devices such as bazookas, ordnance, and duds that they found lying on the ground on Barranca Mesa—now an upscale housing area. Dimas Chavez, whose father moved from Thoreau in central New Mexico to Los Alamos to work for Zia in August 1943, recalls how, in 1946, he and his friends played with an old undetonated bazooka shell, tossing it around, banging on it, and dropping it from the first-floor balcony of the Sundt house they lived in into the garbage can. Sometime later, twelve-year-old Leroy Chavez and Don Marchi, who was only five at the time, mimicked them. When they dropped the shell, it detonated.[17] The shrapnel ripped open young Leroy's abdomen.

In 1949, the police found four boys sailing their boat in radioactive waters beyond the fences. Breaching into restricted fenced areas was a famous game among youngsters.[18] Bill Jette remembers, "There was all sort of mischief to get into." The first rule was to never go beyond the fence, but then he said, "I don't think I was there a week before I went through that fence. . . . And we'd play in the canyons, just explore."[19] One 1948 article displayed a photo showing three boys with their bicycles playing in front of a "DANGER, Contaminated area, Do not enter" sign on the wire fence. The photo's caption read: "Residents get used to signs like this one, and presently are no more alarmed by them than you would be by signs saying, 'Keep off the grass.'"[20] There could not be a better illustration of the safety/security paradox.

Thus, the town combined the dangers of littered ordnances, dumped wastes, and an enemy's threats of nuclear attack, but the population thought of their community as a safe and healthy environment. Secrecy greatly contributed to the endurance of the Los Alamos myth of a dream-like suburban community where residents shared a common purpose to protect the nation. At the same time secrets contributed to the stratification of the community and of the workplace. Anthropologist Hugh Gusterson conducted fieldwork at the Lawrence Livermore National Laboratory to understand the mindset of weapons lab employees and determine how some people came to believe "that the development of nuclear weapons made both superpowers more secure," while others believed that "the stockpiling of nuclear weapons by the superpowers was a terrifying act of lunacy."[21] He analyzes the discriminating capacity of secrecy, writing,

> From Edward Teller's famously frequent retort to critics at public meetings—"If only you knew what I know, but I can't tell you: it's secret"—to the many scientists who told me that antinuclear protestors "just don't understand," the scientists' reflex is often to respond to criticism by claiming privilege. In constructing this sense of privileged status, the rituals of secrecy compound the effects of scientific training at elite universities, where scientists learn a robust confidence in scientific knowledge and a disdain for the superstitious views of the laity. Their standing as scientists who understand the secrets of nature is magnified by their status as scientists who know the secrets of state, so that there is a double sense in which protestors "just don't understand."[22]

Thus, nonscientist employees did not identify discrimination within the nuclear research facilities for what it was because it was constantly associated with knowledge. It seemed normal that scientists be privileged on many levels because they *knew*.

Embracing Nuclearism: Better Us than Someone Else

Secrecy at New Mexico's other defense-related facilities has not been as documented as in the town where the atomic bomb was born. However, they were no strangers to secrecy or risk either. In his second "Hush-hush" article in March 1954, George Fitzpatrick comments, "We quickly discovered that to protect this secret experimentation, security precautions are just as strict at White Sands as at Los Alamos." Fitzpatrick goes on to recount the convoluted path through security procedures that prevented him from even getting "a bird's-eye view" of the launching of a 5,000-miles-an-hour rocket. Like in Los Alamos, the particular atmosphere of the rocket boomtown found its way into children's games: "No wonder that the kids down at WSPG like to play with space ships and ray guns instead of playing cops and robbers. No wonder, when they write to Santa Claus, instead of wanting cowboy suits, they ask for space cadet uniforms!"[23] Within two generations, the change was radical between David McDonald's wish for his first saddle as a child on a ranch in the Tularosa Basin and the high-tech heavy Christmas lists of children on the military reservation.

Living in New Mexico, surrounded by military installations, was riskier than residents could have known. In 1986, the story of a "broken arrow"—an accidental nuclear event involving nuclear weapons, warheads, or components that does not create the risk of mass death—in Albuquerque was reported to the public based on military documents recovered through the Freedom of Information Act of 1966. On May 22, 1957, a B-36 bomber preparing to land at Kirtland AFB accidentally dropped a ten-megaton hydrogen bomb south of the city.[24] The weapon fell on inhabited land, and its explosives—not the nuclear device—detonated, creating a twenty-five-foot-deep crater. In 1994, Professor of History Les Adler commented on the event and on living in New Mexico,

> Those of us living in the region had long known, and, indeed, were strangely proud of the fact that Albuquerque was likely to be a major enemy military target due to the region's role in the production, testing and storage of atomic and hydrogen weaponry. . . . For a town without major league credentials in any other fashion, this fact produced a certain cachet, particularly in an age of bomb shelters, civil defense programs and above-ground testing in nearby Nevada.[25]

The "cachet" Adler talks about resulted from nuclearism. The expression "strangely proud" reminds us of Stanley Kubrick's 1964 film *Dr. Strangelove or: How I Learned to Stop Worrying and Love the Bomb*. The more important the region's role within the national complex, the more nuclearism had to be infused as one of its major "credentials."

Secrecy at Sandia was just as draconian, but it also proved to be a hindrance to the development of its corporate image. The promotion of the new corporation to the local population and the recruiting of staff members were hampered by restrictions on the information that could be publicly released. As a result of limited communication on its operations, Sandia relied on contacts between executives and local "business, civic, and service organizations to promote community cooperation between Sandia Corporation and the City of Albuquerque."[26] These contacts were facilitated by Sandia officials who were active at the City Commission, the Chamber of Commerce, and the Community Chest to deliver speeches on Sandia's role in the atomic-energy program and on their employment needs. Another problem was that many who were recruited would accept other employment during the long delays induced by security clearances.[27] Once the paperwork was finalized, working in the nuclear armament industry could pose a challenge to the personnel who entered the culture and ideology of nuclearism.

Secrecy allowed a larger field for interpretation of one's controversial work, since no information was available beyond one's specific task. Some employees found it morally trying because they considered the escalation of nuclear arsenals inherently wrong, or because they became concerned about the industry's environmental impact as the dangers of radioactivity became more widely known. However, others fully embraced nuclearism and felt pride in their contribution to American military and technological supremacy. This was often expressed in the recurring opinion "better us than someone else." A poll in August 1946 showed that the top three reasons to keep the bomb a secret were that "other countries would use the bomb" (32 percent), that "it is in good hands here" (26 percent), and that "it is a protection for us" (18 percent).[28]

Outside the fences, the population had no other choice but to trust these white-coated soldiers of the Cold War with weapons of genocide. Tad Bartimus comments, "The perception on the outside of a weapons lab can be that of a secret hideaway filled with crazed Dr. Strangelove warmongers hankering to see mushroom clouds. That's not the reality. Soul searching goes on inside the Labs: Are we doing the right thing?"[29] As proof of this "soul-searching," Bartimus relies on interviews of LANL employees who had to justify their work, sometimes to their own families. He thus recounts the story of Allen, Ted, and Hugh Church, the sons of Fermor Church and Peggy Pond, who had worked at LARS before it was taken by the government. When Bartimus interviewed the Churches in the late 1980s, Fermor had become an environmentalist with New Mexico Citizens for Clean Air and Water, and Peggy nourished strong antinuclear sentiments alongside her husband. Their three sons, however, had gotten jobs at Sandia, working with nuclear weapons.

Ted was first to get a job there. He says, "Some folks always had a problem with me and what I do—what they think I do. . . . Friends have a saying though: 'Better to be in the world than out of it.'" His mother "was always upset that the three of [them] ended up together here." He later became a Quaker, but believed his faith was compatible with his work because he felt that it was the scientists' and engineers' "responsibility to help the public understand." After the Korean War, Allen came to work with Ted in the power supply group before becoming a timing expert for nuclear weapons. Allen reports people calling them "crazy people who work with nuclear weapons," but replied to these accusations, "we are not crazy. I find people here to be very sincere; they take it with a great deal of responsibility. I'd rather be a part of it than leave it to someone else." The youngest son, Hugh, joined Sandia in 1957 to work on atmospheric weapons effects. By then, it had already become a four-thousand-employee corporation and was deeply involved in testing operations at the Nevada Site. He recalls, "You wonder why you have to be working with weapons of destruction when there's so much else to do out there. But I believe the deterrence aspect has kept us out of war for so long. You make every effort to make sure things aren't used in the wrong way. And there's comfort in having somebody even-tempered handling it."[30]

Beginning in the 1960s, with the rise of anti-nuclear opinions, the mere association with Los Alamos had become prejudicial. Children raised on the hill encountered negative remarks when they went out of the community and said where they came from. Bill Jette, whose father Eric Jette was a metallurgist during the Manhattan Project recalls, "It seems like I spent half my time defending what they did up there . . . One of my sons is a child of the 1960s. He has a problem with what his grandfather did. We don't even discuss it." From the onset, atomic workers have struggled with the consequences of their work and on how to justify the philosophy behind it.[31] Still today, a certain stigma remains about Los Alamos children who are jokingly said to glow in the dark.

The endurance of nuclearism over seven decades of the Lab's existence shows that it was intrinsic to the culture of the Laboratories. Across the spectrum of employees, all were more or less influenced by it. Florida Martínez says that working at LANL made her proud because she believed that the Lab was there to protect the nation while offering valuable jobs in nice and safe facilities.[32] Likewise, Delfido Fernández was convinced that the Lab would never close "because it's a very important place for the nation."[33] And Paul Emilio Frésquez, who had a security job as a lieutenant at the plutonium facility with a difficult schedule of sixteen-hour shifts, declared, "National security demands me to be there." He and his team would go from room to room and look for security infractions such as documents that would not be locked up for the night. When Frésquez was hired, he was asked to give a written statement about what he thought of the protection of Los Alamos. He stated, "The reason we're up there is to protect the lives and the property of the Lab and there are a lot of lives up there and down here [in the Valley] we are protecting."[34] A "Salute to Los Alamos" in 1955 stressed the two communities' common goal of national security as a bond:

> Los Alamos and the Espanola Valley are closely tied by work, business, entertainment, and personal interests. Many who work in Los Alamos live in the Valley, own property here, and send their children to Valley schools. Los Alamos residents come to Espanola Valley to shop, to attend rodeos, and other celebrations and events. This city of thirteen thousand "on the Hill" continues to be one of the most important cities in the World—and Espanola Valley is proud to salute the Laboratory and the thousands of people who are making history in the great atomic research projects of our government.[35]

Still, there remained the fact that lower-skilled employees did not have access to the same amount of information as staff members, and as a result of this disparity, their sentiments toward their workplace depended primarily on how well they were treated and sometimes what they heard or read about the Lab. For the workers who were aware of the risks involved in their jobs, the specter of unemployment often made them resort to willful ignorance, turning a blind eye to the possibility of health and environmental damage. That is why, despite scandals about the state's nuclear facilities in the media, there was generally little opposition to the nuclear weapons complex, which had become indispensable to the local economy.

Furthermore, the ideology of nuclearism is closely intersected with patriotism. Many New Mexicans expressed their pride in taking part in the ideological strife against the Soviet Union. The atmosphere of competition in the state peaked every time a gap was suggested between the two superpowers and revived the original spirit of the Manhattan Project. Major Brown, who presented the Alamogordo Guided Missile Test Base in 1948, heralded, "It is felt that New Mexico will be justly proud for having had this organization operating within the State, whose basic discoveries will be affecting the history of man for many generations to come."[36] In Albuquerque, efforts concentrated on education. The call for more focus on hard sciences was a nationwide phenomenon in reaction to the Soviet launch of Sputnik in 1957; and the atomic city was to be a good student.[37] In 1959, the University of New Mexico received a $227,916 grant from the AEC to establish a laboratory for students pursuing a science degree with a major in nuclear engineering.[38] A year later, the State School Board amended the graduation requirements for the city's high schools to include more mathematics and science courses.[39] Other measures included the development of competitive grading systems in elementary schools, augmenting faculty salaries at the University of New Mexico, and offering bigger prizes at the New Mexico Science Fair.

At the beginning of the 1990s, nuclear weapons technology still accounted for more than half of the work at LANL, and scientists still faced the same moral dilemmas of nuclearism. Jo Ann Shroyer addresses some of these longstanding predicaments and humorously disputes the reputation of Los Alamos as "the world's supply of nerds—badly dressed, absent-minded brainiacs who can't manage to wear matching socks" to describe the lab as a "complex," "not easily understood" place, "a community that has been both a beneficiary and a victim of its own history and mythology."[40] One of her interviewees, Jas Mercer-Smith, reiterated the nuclearist ideology, which had not changed since the 1950s. He compared Lab scientists to fairytale witches, saying, "Our job is to scare little children into behaving" and referred to the paradox of his profession, which is "to guarantee peace by making weapons so horrible that no country would ever be foolish enough to provoke their use," in other words, *si vis pacem para bellum* ("If you want peace, prepare for war").[41] After admitting to having felt the adrenaline rush at countdown and the euphoria of a successful shot at the Nevada Test Site, Mercer-Smith added that their vocation was "much more like art than most science."

New generations of Lab scientists, however, sought to detach themselves from previous practices. Tom Ribe, another interviewee, said he viewed the AEC as "almost a criminal agency in terms of what they were doing to the people [soldiers and downwinders] in the 1950s and '60s."[42] Ribe's opinion touches on the most dangerous, ethically, and morally reprehensible secret of the nuclear era, which has affected soldiers, downwinders, uranium miners, laboratory workers, and the general public: radioactivity.

Standardized Secrets

New Mexicans still struggle with the aftermath of the dangerous practices of World War II and the Cold War, many of which were used without their knowledge. Originally, the atom was perceived in the light of promising prospects. Ideas such as atomic cars and artificial suns—"nuclear utopia" in the words of Paul Boyer—were popular in the postwar years. Medical applications seemed promising: *Operation Atomic Vision*, a high-school study unit published in 1948 by the National Education Association, declared that atomic energy would reduce the likeliness for that year's high school students of dying prematurely of cancer, heart disease, contagious diseases, or any other afflictions of the time.[43]

The powerful AEC was then both responsible for developing the nuclear weapons complex and for regulating it, that is to say, for protecting Americans from the hazards produced by the same complex. It took until the 1974 Energy Reorganization Act, which split the AEC into two agencies (the Department of Energy, in charge of development, and the Nuclear Regulatory Commission, in charge of protecting public health and safety) to restore the principle of checks and balances in nuclear decision making. While the AEC endeavored to portray atomic energy as positive, it also sought to allay fears of radiation. Secrecy was the method of choice to ensure that the public could not discuss the hazards of atomic research, weapons testing, and stockpiling. In addition, the hazardous aspects of bomb making went unnoticed as long as anxiety about Soviet bombs was the central concern. Once that changed, the AEC presented itself as being in control of what they considered were safe dosages.

The problem was that dosages had never ceased to decrease. Initially, the very first limits were set at the appearance of observable signs of excessive exposure to radiation, such as skin ulcerations, whereas now the standards are based on the prevention of exposure-related diseases or stochastic effects, such as cancers and genetic mutations. Originally, scientists believed only a high dose of ionizing radiation would be dangerous. By 1903, the recommended limit was set at 10 rad (absorbed radiation dose) per *day* even though animal studies had already shown "that x-rays could produce cancer and kill living tissue." In the 1920s, the recommended "toleration dose" for workers had gone down to about 70 rad per *year* after an investigation conducted by a medical examiner linked the ingestion of radium by watch-dial painters to serious disease and death.[44] By 1934, the dosage considered safe for humans was down to 36.5 roentgens (more or less equivalent to rads) per year and 5 roentgens per day for individuals who were occupationally exposed to radiation. A one-time 5-roentgen exposure was also the limit after the Trinity blast, a level which is now considered dangerous. It became the *annual* limit recommended by the International Commission on Radiological Protection (ICRP) in 1957 with the discovery of radiation-induced leukemia. The measuring unit is now the REM (Roentgen Equivalent Man) and the millirem, which measure the biological impact of roentgens on humans.

We now know that plutonium lodges itself in the bone marrow and can quickly affect all parts of the human body. The isotope will remain hazardous and continue to irradiate and damage surrounding tissue as long as it remains radioactive. With a half-life of 24,000 years, exposure will entail consequences including an increased likelihood of cancer, cell mutation, and interference with blood components. At the time of the Manhattan Project, however, the full effects of radioactivity were still largely in the realm of speculations. The Met Lab program headed by the University of California's most prominent radiologist, Dr. Robert S. Stone, was initially established to determine a threshold for Manhattan Project employees. Then, General Groves decided to build a separate military medical program within the District, the Radiological Division, headed by Stafford Warren. The military staff of this program owed "unconditional allegiance to the military-industrial culture and to Groves" so that the doctors working on "unique occupational hazards" would not hinder the continuation of the Project.

In August 1944, a Los Alamos chemist breathed in and swallowed the content of a vial of plutonium that had exploded in his face. The doctors called Chicago, but the Met Lab Health Division could not answer their questions. Hempelmann and Oppenheimer pressured Groves into accelerating a medical research program at Los Alamos, but it never came to be. The accident precipitated a confrontation between Warren and Stone, whose opinions illustrated the two conflicting trends regarding health standards: purposeful ignorance to avoid slowing down the project on the one hand, and research to eliminate hazards altogether on the other. When patients presented signs of radiation illness doctors could not make a clear diagnosis and had little idea of how to treat the symptoms. Radiation poisoning was referenced as "medico-legal complications" on patients' files and they were told that even though their disease was not District-related, District doctors were best able to treat them.[45]

In the late 1940s, the atomic bomb survivors were closely monitored by American medical groups. Colonel Stafford Warren headed the Manhattan Project Atomic Bomb Investigation group in Japan. They recorded witnesses and helped write a report titled *The Effects of the Atomic Bombs on Hiroshima and Nagasaki*.[46] These studies enabled US doctors to gather information on how radiation from an atomic blast affects the human body. Around the same time, medical experiments were conducted within the Project on individuals, who were given small doses of radioactive iodine, tritium, and other substances to see how fast they would expel them. In 1973, an inquiry requested by the AEC to determine "whether informed consent was obtained from patients injected with plutonium" concluded, "Formalized standards for patient consent to experimental procedures did not exist prior to 1946" and "security considerations could have interfered with whatever disclosure the investigators in these plutonium studies may have considered at the time. The word plutonium was classified until the end of the war."[47] Such conclusions demonstrate again how national security superseded moral concerns through the means of secrecy. It took until the 1990s for these experiments to become public knowledge when Energy Secretary Hazel O'Leary made the decision of disclosing documents on human radiation testing on retarded children, terminally ill patients, pregnant women, and prisoners.[48] Eileen Welsome compiled these stories in *The Plutonium Files* in 1999.[49]

The risks and experiments of the 1950s—including atmospheric testing, the use of "atomic soldiers" as guinea pigs and exposing populations around test sites to radioactive fallout[50]—were made possible by the threshold theory that postulated that radiation would have no permanent effects if the exposure remained below a minimal level and only accidents could be fatal. In 1953, the AEC proclaimed, "Low levels of radiation produce no detectable somatic effect; that is, the body is able to repair the damage virtually as quickly as it occurs."[51] In 1956, however, Dr. Alice Stewart published her pioneering study on the damages abdominal x-rays, even in a minute amount, could cause to the fetus of a pregnant woman. Stewart's research shows a clear connection between childhood leukemia and x-rays in early pregnancy, which greatly challenged the threshold theory.[52] In the early 1960s, concerns arose when elevated rates for leukemia and solid-tumor cancer among atomic-bomb survivors further "brought into focus the possibility that even low levels of exposure might induce cancer."[53] Nevertheless, a new limit remained difficult to determine.

Eventually, in 1977, Los Alamos was subjected to two different regulations: one established by the EPA for public exposure outside the Lab property set at 25 millirems per year and one established by the DOE and NRC for public exposure at the Lab boundary and outlying areas set at 170 millirems per year and 500 millirems for people who live adjacent to the fence line of nuclear installation. The maximum permissible radiation dose for nuclear workers remained set at 5,000 millirems (or 5 rems) per year. Supporters of the industry argued that the decrease in standards was due to technological progress that made such reductions more practical rather than due to new evidence of biological damage from low levels of radiation.[54] The following decade saw the intensification of the anti-nuclear movement in the country. By 1993, the National Council on Radiation Protection and Measurements (NCRP) recommended an occupational dose limit more than three times lower—1.5 rem per year over a working lifetime of forty-seven years.

Setting radiation safety standards was a long and difficult process as secrecy prevented public scrutiny, and nuclearism often influenced scientific findings in order to comply with the desires of the supporters of the nuclear weapons complex. Gerald Clarfield's conclusion to his work on *Nuclear*

America in 1984 was commensurate to the scandal produced by the news of the long-term impacts of atmospheric testing. He writes,

> American officials consistently adopted the most optimistic view of risks possible; they denied reality and common sense; they placed an impossible burden of proof on their critics; they shut their eyes to scientific evidence that they should have considered; they manipulated data to produce results they wanted; they tried to repress investigations that they feared would produce unwelcomed news. . . . The most that we can say now, as historians working exclusively from printed records, is that they acted with extreme and willful carelessness, with a monstrous disregard for the possible sufferings of innocent people.[55]

Among these innocent people were the New Mexican workers employed at the state's nuclear-related facilities. Many have witnessed these changes in safety regulations, the medical checkups, the influence of nuclearism ideology, and the eventual revelations on how much risk they had been taking on the job firsthand.

Hazardous and Extra-Hazardous Tasks

Risk taking was promoted and rewarded on the paychecks of Zia employees who accepted tasks in the Tech Area classified as "hazardous" or "extra-hazardous" by the University of California. A third of Zia employees worked in the Tech Area. An employee could increase his pay by 10 percent for hazardous jobs or even double it for extra-hazardous jobs. These jobs included cleaning material under the supervision of California experts with Geiger counters. Their work clothes would be "laundered after work at a University of California–operated laundry that cleanses the wearing apparel of radioactivity and tests them before they are worn again." According to McKee, the head of company, most of these employees were "completely accustomed to" the safety measures, and "those drawing double pay for extra-hazardous jobs" would "rarely willingly consider a change, nor are they injured by their work." He adds, "Thus far no Zia employee has been either gravely burned or become seriously radioactive, which is a tribute to the safety measures of the University of California."[56]

The impoverished population of northern New Mexico was prepared to accept working hazards because many of the jobs they had had access to in the past—in the mines and on the railroad—had been just as dangerous. Josefita Velarde recalls a conversation she had with a nurse when she started working in Los Alamos in February 1946, asking her whether she knew that the atomic bomb had been made here. The nurse, who was from up-state New York, replied, "What? They made it here? I'm leaving!" Yet, Josefita could not recall any of the local people saying that they were scared because no matter where they worked, it had always been hazardous for the working class. Before the war, people also succumbed to tuberculosis and influenza. The only safe place were the ranches where people survived because of the feeling they had for the land, "a real mystical pull." But she strongly believed in the Lab as a lifesaver. So, Josefita concludes, "Alright, they made the bomb, who cares? It was a job, good paying jobs! The whole state is affected."[57]

Years later, though, some of these former employees described their work conditions in the 1950s in a different light. One of them was Genaro Martínez, janitor at Los Alamos for twenty-five years. He worked in contaminated areas with a respirator all day long and confirmed that decontamination was a different classification that would increase their pay by ten cents an hour. Machinists would test tubes or pipes to know if they were "hot"; they would mark them with chalk or tape, and then the janitors would wash them until they were "cool." For protection, they would be given clothes and rubber gloves. Genaro often tested himself to check contamination levels. He recounts the story of a colleague whose hands had gotten contaminated, and, as a result, had to wear gloves to eat. Worried, Genaro went back to regular cleaning, but this angered his Anglo supervisor who transferred him to night shifts in the S site where water used to run beneath the buildings; Genaro hypothesized that it was contaminated. In 1958, he initiated the creation of a union for janitors because of the "nasty" treatment they had from supervisors. There were about three hundred janitors and, according to him, they were the Lab employees with the lowest pay.[58]

The consequences of these extra-hazardous tasks only began to show years later when workers were confronted with health issues that could be linked to overexposure to radioactive materials. Rubén Montoya, who had started working for the Lab in the late 1940s, was advised to take medical retirement in 1972 after hurting his back on the job. When he was doing research on the joining of beryllium with soft metals, the medical team would check his breathing and take x-rays of his chest every three months. The letters read "no abnormalities detected." In 1978, however, after a chest x-ray at the Veterans' Hospital, he was told that he had an enlarged thyroid. In his interview, he described his job working with dangerous materials, saying, "We had uranium all over. We had a hazardous materials book. We would be working under a hood, we used a respirator. We took all the precautions we could when I recognized we were working with dangerous material. My boss didn't like that because it took too much time. . . . I liked what I was doing, but *I didn't like what they were doing to me.*"[59] Ruben's discourse opposing "I" to "they" clearly expresses the sentiment of being exploited and the dilemma of appreciating one's job at the same time.

In the 1970s, a lawsuit charged LASL with damages incurred as a result of working with radioactive materials.[60] It was filed by Ramon Martínez, fifty-seven, of Española, who claimed he had been "disabled by a 'neurotic fear of radiation' caused by his work in a LASL uranium foundry." A few months before his planned retirement in February 1976, Martínez underwent surgery to remove a cancerous tumor in his right eye. Robert Salazar, a colleague, remembered he had left the Lab in 1956 as a result of being afraid of radioactivity. In his testimony, Salazar talked about spills, poor ventilation, the absence of exhaust fans around furnaces, and respirators that did little to prevent inhalation of vapors and gases. He reported that "about 10 to 15 times in the course of his employment in the building, radioactive materials including uranium-235 would 'spurt out' from a centrifugal furnace. 'It would throw it just all over the walls, so you have to go and scrape it and pull it out and clean the rest with acetone.'" Salazar also testified that when the men took off their masks after getting out of the furnaces, "you could see the blackness of the oxides all over your face." Alex Lovato, another retired foundry worker, went blind. Part of his job was to look into the furnaces "where uranium was fabricated with an optical pyrometer." He explained that workers usually wore masks instead of full-face respirators and, as a result, had their eyes exposed. The New Mexico Supreme Court ruled in 1979 that the protective articles were ineffective. Martínez was awarded $75,000.[61]

Other workers, however, believed that there was a good side to secrecy. Jasper Tucker, who moved to New Mexico from Oklahoma in 1954 and worked in road construction and as a heavy equipment operator, comments,

"It was better not to know what was going on. You went in, did your work and didn't ask any questions. Because it pays well, you just have to go with it. If you don't, you're out."[62] The prevailing lack of knowledge and understanding encouraged locals to go on with their tasks without asking for more information. José Benito Montoya mentioned that his parents never got a television because they had heard so much about radiation that they believed a television would give them radiation.[63] This anecdote is proof of how little the first generations of workers could grasp such complex, health-related matters as a result of omnipresent secrecy. Many, like Pedro Martínez, did not believe their job was dangerous. At the time, the safety personnel would give the ok for no radiation, but they found out later that the radiation level was much higher than they originally had thought.[64]

Several interviewees referred to serious accidents. Ramón Frésquez who worked with explosives mentioned six casualties when they were transporting material to be burnt and two people who died in a drilling process on one of the charges—the drill heated up and exploded. Operators were protected against explosions, but accidents still happened.[65] Another worker, Alfonso Mascarenes, had participated in many dangerous experiments, but one time, he said he accidently created a "hydrogen bomb" with flames that went up "75 feet high." He and his colleague came out unscathed. Following this accident with lithium hydride, he became aware of the dangers of treating the substance and turning it into lye. He became more active in chemical waste management.[66] Not all employees were as lucky, and mortal accidents were kept a secret. In 1959, for example, "an explosion at the lab took the lives of four *paisanos*, from Chimayó, Española, and El Guique, but little was heard as to what really took place."[67]

Despite incidents, local businessman Richard Cook affirmed that radiation was not on anybody's mind and strongly downplayed its dangers. He comments,

> Every ten years there would be an accident of some sort, a guy would get blown up, but we worked all over that area, there were sites everywhere but we never gave a thought to radiation. We weren't told there was any radiation and I think it was completely blown out of proportions, really. Environmental issues have been blown out of proportion, and everybody is afraid up there that they are going to be blamed for something. If it was so dangerous up

there, don't you think that people would move? I am not at all afraid
for my kids and grandkids because of the plutonium, etc. I think
there is a certain amount a sacrifice that has to be made for progress.
. . . I admire them a great deal because they are a smart group of
people and to have that concentration of brains and under-utilize it
would be a crime. Somebody ought to take it in hands. They would
say that some places, nuclear dumps were dangerous and we should
stay away, but we would drive by them all the time and it never
bothered anybody. There is no question that there is contamination,
but look at all the advances that were made. You look at the good
and you put up with a little of the bad. They have been so careful up
there now that I don't think there is any possibility of contamination
getting out of hand in the future.[68]

Cook's use of the words "blown up" and then "blown out of proportion" draws a parallel between two kinds of excess: the excessive risks that
could lead to being "blown up" and the excessive reactions that were pro-
portional to the amount of anxiety-provoking revelations the public was
exposed to, especially in the 1980s. The notions of control, to "take it in
hands," as well as that of sacrifice show a full internalization of the nuclearist discourse. For the sake of "progress," sacrifices must be made. For
New Mexicans, however, "progress" did not necessarily mean scientific or
technological progress, but, rather, economic progress out of chronic poverty. Therefore, according to Cook's testimony, most locals were content
to leave the control of their economic fate in the hands of "a smart group
of people."

As a result of secrecy, training was not thorough when it came to
sensitive materials. Despite the high concentration of plutonium at TA55
where he worked as a security guard, Paul Emilio Frésquez felt secure in his
workplace. He believed all the same that his personnel were not trained as
much as they ought to have been on the matter of radiation. Their training
was minimal. He made the connection with secrecy and accountability:
"There's always the question in the back of your mind: are they being
legitimate with me? Are they telling the whole truth? It scares you, makes
you wonder if they are being upfront with you." Sometimes, he would sit
down in a room to have lunch and a crew would come to tell them, "Hey,
no one should be here without a full face mask and a full body suit, this

room should be closed." However, like Richard Cook, he considered Los Alamos to be an acceptable risk, concluding, "We are blessed in this area. We're the envy of the world."[69]

Likewise, Larry Dillion of El Rancho, who was a seasonal heavy equipment operator at Los Alamos between 1991 and 1993, expressed the same unwavering faith in the Lab's research. One of his friends, Doug, tore into an old sewer line in 1991. No one knew it was there, but it contained acid and radioactive material that splashed all over him. He had skin bleeding and could no longer go out in the sun, but, as far as Dillion was aware, he was not compensated. This event did not tarnish Larry's opinion of the Lab, which he sees as a research center with endless possibilities to cure diseases and other problems of modern society.[70] Thus, no matter how risky their tasks or their work conditions were, some workers had such faith in the positive impact of Los Alamos that they came to have a one-sided vision of their experience there, blocking out the more questionable aspects. The very nature of radiation as it is invisible and inodorous, and yet ubiquitous, aided the lack of concern regarding its risk.

Finally, Charles Montaño's testimony is particularly poignant, for it addresses some of the issues that most affected the life of local workers at Los Alamos. His father, who worked for a contractor, used to complain about the incoming Anglos who were given management positions; he would say, "*Estos gringos*, they come from God knows where and they are put in charge." This feeling of injustice developed into "a pain, an anger, a common bonding" among local workers who had the impression that "they were treated like slaves." After going to college, Charles was hired as an accountant in the Nuclear Material Safeguards Department where they kept track of nuclear materials. At first, he recalls being caught up in the "mystique" of being a Lab employee. Then a new generation came in to advance their individual careers without a project or a focus; the mystique and sense of common purpose of the Manhattan Project era had faded. Politics ruled the day; biases became more apparent. On nuclear materials, he professed, "The Lab tried to avoid exposing the Lab, not necessarily the workers, to the adverse publicity of having a bad accident." Accidents still happened, but "they have a public affairs office that can put a spin on virtually anything and avoid the general public from knowing the extent or the magnitude of the problem, whatever it might be." He described the Lab's policy as "a corporate culture which does not want public accountability or scrutiny, that will turn on its own workforce and its members that question the decisions being made by management in an intuitive way."[71]

Charles Montaño's experience reflects that of other New Mexicans who came to understand the downsides of the nuclear bonanza. Some of these workers' testimonies address issues that blended secrecy with security but also with discrimination and exploitation. Since most of the support staff was from the Valley, and dangerous tasks were assigned to them, the correlation between origins (or, rather, the education level that was also related to origins) and exposure to danger is clear. This is one of the components of a neocolonial labor system in which the local, poorer population is assigned the riskier tasks. Montaño's description of the feeling of inferiority by referring to slavery is extreme, but it confirms the sentiment of being conquered and exploited. His generation experienced the disillusionment of being trapped within a discriminatory framework. Former generations had come out of poverty unconcerned with the jobs' shortcomings, for it was a formidable improvement in comparison to their previous struggles. Younger generations, however, became more critical as secrecy was lifted on certain alarming, inequitable, and unethical practices.

Nine

Dangerous Practices, Toxic Legacies

"With a collective memory that tended to recall the political and financial rewards of past expansions, rather than the societal and environmental costs, the nation entered the Atomic Age with little to guide it except a deeply held belief in progress."[1] This remark by journalist Kevin Fernlund underlines how the American West became "the front line" in the Cold War in the same way it had enabled the United States to become an economic superpower in the previous century. Many turned a blind eye to the adverse effects of building the nuclear arsenal, which shows, as Fernlund notes, the tendency of collective memory to be oblivious to the costs of conquest, especially when they impact animals and plants before humans.

Waking Up: From Secrets to Scandals

In New Mexico, the realization that the environmental impact of the weapons complex could have frightening long-term repercussions occurred at the turn of the 1970s into the 1980s with more signs of distrust shown toward actors of the nuclear industry. Not long after leaving Harrisburg, Pennsylvania, the site of the Three Mile Island accident, to settle in Taos, Allan Richards wrote in the *Taos Magazine*, "The public is on the edge. It has learned to fear. It lives in a cloak of mistrust. Weaned on paranoia, it knows the government doesn't tell the truth, the whole truth . . . but what it sees fit to."[2] With this realization came a new era of conflicting interests and major trust issues between the New Mexican population and those at the head of the state's nuclear facilities. It seemed that one cloak (of secrecy) was replaced by another (of mistrust). Militant and watchdog groups mobilized against officials who sought to minimize the environmental and health impacts of the complex, and the public became more systematically confronted with two versions of the story: the alarming tale of experts commissioned by activists and the reassuring stance of state and corporate authorities.

The confrontation of numerous viewpoints and a widespread will to maintain the state's prime source of jobs resulted in the lack of powerful, united protest against waste disposal, leakages, contaminations, accidents, and cancer rates. Fear of nuclear war had led the population to accept the risks of building an exorbitant stockpile of weapons, yet many took those risks for economic reasons that had little to do with the balance of power. Rather than attributing blame or victimizing any part of the population, my point is to address the issues that emerged when factoring in the environmental consequences of the nuclear buildup. Furthermore, since other authors have admirably addressed the environmental legacy of the Manhattan Project in New Mexico,[3] I will not attempt to recapitulate all environmental and health concerns within the state but will focus on the central issues that New Mexicans confronted, which illustrate the transition from blind confidence to mistrust and relate to nuclear neocolonialism. The well-oiled economics and politics of the nuclear weapons complex were disturbed by uncomfortable truths, and all who benefited from the industry were facing the uglier profile of the nuclear golden goose. Someone had to be found responsible for the damage.

After the Chernobyl accident in 1986, Congress ordered the Secretary of Energy for the National Research Council to investigate health, safety, and environmental issues arising throughout the American nuclear weapons complex. Chairman Richard A. Meserve notes, "Acceptable risk must ultimately be measured by balancing the benefits of the activities against their costs."[4] Expressions such as the "burden of past operations" appear repeatedly in the text, clearly laying the blame on past actors and past decisions for the "potentially serious environmental problems" that now plague the complex. As a result, "substantial pressure has arisen to restore the environment—a prospect that is both technically and financially challenging."[5] Difficulties included the fact that the Department operated under much more public scrutiny and that it was no longer independent as other agencies had entered the equation. Safety and environmental standards had become more stringent, the budget had been cut, and, although nuclear weapons remained necessary for national security, they had lost some of their significance. In other terms, the costs had begun to exceed the benefits. The public's growing awareness concentrated on responsibility and accountability because those responsible for the damage had to be put in charge of the cleanup operations.

In the National Research Council's report, the Department is advised "to learn from its past mistakes" by "improving current waste management practices and developing innovative waste management technologies." Dangerous practices were born in the hectic wartime atmosphere of the Manhattan Project: contaminated material and chemicals were buried in cardboard boxes sealed with masking tape before being placed in unlined and unprotected trenches no matter their radioactive composition; gases were vented to the atmosphere after limited filtration; and untreated, highly contaminated liquids and solvents were kept in unlined ponds and flushed into the surrounding canyons.[6] A 1962 inspection found that one gate left open had led to a radioactive waste pit becoming a junkyard. In 1975, the US Energy Research and Development Administration (ERDA), successor to the AEC, discovered a plutonium "pocket" south of the Los Alamos Inn. This was originally the location of the Technical Area laundry where the workers' clothes contaminated with plutonium would be washed.[7]

Few records of these events and of the various burial sites' locations were kept, so the Lab had trouble finding all its nuclear dumping grounds and did not know what had been buried in them. Interviewee Ramón Frésquez talks about how the Lab s attitude had changed, saying, "They would just tell us what to do and where to deliver the material—they wouldn't tell us if a site or material was dangerous for the environment. They became more stringent and more respectful of the environment in the last few years."[8] Paul Emilio Frésquez also comments on the pervasiveness of dangerous substances: "You're always wondering if they're releasing something into the ground or something like that detrimental to all of us."[9] While disposal practices improved in other industries, improvement was significantly slowed down by the absence of public scrutiny at weapons facilities. Cold War secrecy was thus the designated culprit of this perilous legacy. Since the 1970s, however, the blame has been distributed to many others—the AEC, the DOE, the Labs, the uranium companies, the State

Environment Department—and many believe the problem of dangerous practices is not a thing of the past.

Toward the end of the Cold War, different institutions undertook environmental investigations to evaluate the extent of the impact of past practices on New Mexico's environment. Parallel to the actions of scientists and federal officials, activists and Pueblo representatives pushed for independent studies. The alarm was first sounded concerning waste management. LASL reports as early as 1973 collected evidence of radioactive contamination of the flora and fauna in the Mortandad Canyon liquid waste disposal area.[10] As part of the study, scientists placed bee hives near the point where the liquid effluents were discharged into the canyons. Tritium was found to be the greatest source of contamination among the worker bees. Another study on fruit and vegetables revealed the presence of radioactive contaminants, including uranium and plutonium, which LASL scientists believed were likely due to fallout from nuclear tests.

The Love Canal scandal of 1978 and the Three Mile Island accident a year later added fuel to the fire. In 1980, Phil Niklaus and Dede Feldman published a booklet that gathered a series of articles on the handling and disposal of nuclear waste at LASL in the *Albuquerque Journal*. As part of their investigation, Niklaus and Feldman interviewed some forty individuals including current and former LASL scientists, state and federal health and environmental officials, state legislators, and Los Alamos area residents. Their findings were irrevocably bitter. They write,

> We have come to believe that the people of New Mexico have been seriously deceived as to the nature and extent of the routine and accidental releases of radioactivity from the lab. LASL, born in secrecy dictated by wartime conditions, has never been able to shake the habit. Information on lab activities, channeled through the New Mexico news media, continues to be seriously tainted by half-truths, routine down-playing of radiation accidents and, in some cases, outright falsehoods. As a result, the people of New Mexico have been lulled into complacency over the nuclear projects underway on "the Hill."[11]

Secrets were again branded as the culprits and became the main battleground for the fight between the labs and journalists. The public was caught in the middle. The release of chemical and radioactive substances into the environment was the most concerning issue cited, even though it was denied in LASL reports distributed to the media. A few LASL officials criticized local newspapers for arousing the public's fear of radiation by not complying with the Lab's reports. The DOE and LASL also invoked a lack of public interest to justify why there had been no public hearing upon the release of their final Environmental Impact Statement on LASL operations, contrary to other DOE sites such as the Lawrence Livermore Laboratory. An editorial cited LASL's annual budget as an "example of the arrogance the industry has toward the public. . . . Less than 1 percent—approximately $1 million—is budgeted for the disposal and monitoring of radioactive waste at Los Alamos. Scientists believe the amount is adequate. The public naturally is skeptical."[12] Two years later, the Lab spent $2.8 million and employed a staff of fifty-five people to handle its waste. According to Tom Kennan, head of the Waste Management Division, no chances were being taken. There were no Love Canals in Los Alamos.[13]

And yet, as early as the late 1960s, the AEC already had reclaimed some areas that had been polluted by the haste, ignorance, and thoughtlessness of previous generations. These first clean-up operations, however, had been carried out in secret. Perhaps the most well-known cleanup site, to which much attention was drawn over the years, is Acid/Pueblo Canyon. The Laboratory used it until 1964 as an initial long-term dumping area for liquid waste. The 25,000-gallon-a-day radioactive liquid waste generated at the Lab underwent processing at two treatment plants, where the most hazardous concentrations of some radionuclides were separated as sludge for burial in one of the Lab's solid waste storage sites, while the remaining, still partially contaminated, liquid effluent was released into the canyons. Cleanup started in 1967 and took two years, during which about six hundred dump-truck loads of dirt and debris were removed. Dr. Wayne Hansen, head of LASL's Environmental Surveillance Group, says the radiation from plutonium found in the soil samples at Acid Canyon was "below any applicable standards." He admits, however, that if they had been absolutely unconcerned, his group would not have existed. People who hiked in Acid Canyon were given no warning about the radiation, but, according to Hansen, there was no reason to inform them as there was no hazard, and he professes he would readily have taken his children to the Canyon.[14]

Peggy Franklin and Shirley Walkup played in Acid Canyon next to their houses as children. In 1991, Walkup died of a cancerous brain tumor at age fifty-five. Franklin had successfully recovered from a brain tumor at age fifty-one after having already had a tumor removed from her knee at age fifteen. They both blamed the stream down in the Canyon for their health issues. Declassified reports showed that from 1946 until 1951, the pipe was spewing 10,000 gallons of water a day contaminated with raw plutonium, uranium, and tritium. Then, from 1951 to 1964, even if the waste was treated, it still exceeded modern standards for safety.[15] In 2008, when faced with a wrongful-death lawsuit, the Regents of University of California officially acknowledged previous practices of discharging toxic liquids into the canyons.[16] The suit was brought forth by the daughter of Lowell Ryman, who had died of cancer in 2005 at age sixty-three. The family contended that exposure to radioactive wastes while playing in canyons as a boy in the 1950s led him to develop multiple myeloma as an adult.[17] Two women, who had developed thyroid cancer, joined in. The costs forced the plaintiffs to drop the case.[18] In many instances it took legal action to prompt the laboratories and other responsible entities to acknowledge the impact of their activities upon New Mexico's environment. Lawsuits became a way for New Mexicans to get compensations but also a vehicle for information on the dangers of the industry.

Suspicion grew throughout the 1980s and 1990s. In 1989, the EPA identified more than six hundred sites that needed to be cleaned up at Los Alamos. Most were on Lab property, but some were downtown, close to residential areas. Two years later, *The New Mexican* published the results of another three-month investigation identifying more than 1,800 hazardous waste dump sites to be cleaned up. One controversy at the time centered on the issuing of a permit for the Labs to use a waste incinerator.[19] C. Kelley Crossman, supervisor of the Hazardous Waste Bureau at the Environmental Division, drew heat from the public when he announced he was satisfied that radioactive incineration met federal regulations. Bradley Hanson of La Madera asked Crossman, "How in all honesty can you recommend this permit for approval when, in all honesty, you have neither the staff nor the funding to make sure Los Alamos Labs, and the people they are working for, are telling you the truth?" Franck Walker of Ojo Sarco asked, "How do I know that all the compounds are tested for and some of them don't fall through the cracks—that some of this stuff doesn't become an airborne Love Canal?"[20] These reactions are demonstrative of the distrustful mood toward the Laboratories in a national context of anxiety regarding the impact of hazardous industries on environment and health. They also reflect the anger and growing fear among the population of developing diseases after being exposed to a seemingly harmless and yet noxious environment.

Nuclear Paranoia

Despite the reassuring discourse of authorities, citizens grew warier, developing worries verging on paranoia. Anthropologist Joseph Masco names this phenomenon the "nuclear uncanny," a concept he describes as a colonization of psychic spaces by fear of radioactive contamination, leaving people "to wonder if invisible, life-threatening forces intrude upon daily life, bringing cancer, mutation, or death." As a result of this paranoia, people living close to nuclear facilities have often come to see radiation as "a means of explaining all manner of illness and misfortune—its very invisibility allowing its proliferation in the realm of the imagination."[21] Fear, which had originally been fostered by the AEC to guarantee public support in the arms race, turned against the complex as paranoia shifted from fear of outside enemies to fear of their own backyard.

Housing built on contaminated ground became a source of concern for residents on the Hill. One nuclear test site for high explosives, Bayo Canyon, was opened to housing in the 1980s after cleanup was completed—a 1976 survey had found three times the normal level of radioactive strontium in the soil.[22] Today, housing extends to Barranca Mesa. The proximity between private houses and cleanup sites aroused speculations on the effect they could have on the residents' health. The incidence of cancer in the area started to draw attention in the 1970s. One New Mexico Tumor Registry study indicated "that cancer mortality rates in Los Alamos County in white males from 1950–1969 ranked highest compared to control counties for leukemia, lymphosarcoma, cancers of the liver, prostate, and bladder." Between 1969 and 1974, the incidence of breast cancer in white females was more than twice the US average and several cancers in males—large intestine, stomach, pancreas, and rectum—were more than two or three times the New Mexico average. Harold Agnew, Lab Director at the time, "placed the blame for that on the rich foods consumed by the affluent Los Alamos population, including the widespread appetite for hot green chili."[23] Yet, physicians actually considered chilies good food for the intestine. The head of the Health Division contended that the statistics were not significant because they excluded all the other causes of cancer besides radioactivity.[24] Yet, later studies continued to produce statistics above state and national rates.[25]

In most cases, it has proven complicated to establish a link between cancers and nuclear facilities because many cancer-inducing factors have been uncovered. Nevertheless, some accused the Labs of not giving serious thought to radiation as a cancer-causing factor and hiding behind other explanations. Despite the reassurance professed by Lab officials over the years, statistical analyses continue to raise doubts and frighten New Mexicans. Betty Marchi Schulte, whose father was in charge of the Fuller Lodge dining room, was interviewed by Katrina Mason in the summer of 1992. Mason explains, "Her mother had died of cancer, her father had just died while battling cancer, her sister had been diagnosed with cancer, and her husband, who was suffering from cancer, had just had a stroke." Betty voiced her troubling thoughts because "she couldn't help but wonder about the connection of all this cancer to Los Alamos—and worry about whom it might strike next."[26]

For the public, the main difficulties remained the availability and complexity of information, contradictory stances, and the multiplication of coincidental statistics. But the most worrisome, paranoia-producing aspect of living close to nuclear sites is that radioactivity can quickly travel far, especially when transported by the water current or the wind. In June 1988, the Radioactive Waste Campaign, a public interest group based in New York, reported that plutonium was reaching the Rio Grande and flowing downstream as the radioactive waste stored in LANL s fourteen known disposal sites in the canyons spread through erosion, runoff, wind, and dust storms. These emanations endangered Rio Grande water users and close-by communities such as San Ildefonso, White Rock, and Pajarito Acres. Water is a scarce—and for some sacred—resource that has been closely monitored by the region's inhabitants for centuries. New Mexico has six moderately dependable rivers and all human activity in the state is connected to the abundance or scarcity of water.[27] According to the study, there had been "a pattern of growing mismanagement by the Department, which is allowing radioactivity to leak out of the sites through soil, water, and air—in many cases *intentionally.*" The group sent a letter to Congress calling for the immediate end of dangerous dumping practices, urging that these sites be given top priority for cleanup, and calling for the creation of an independent agency to oversee the weapons plants. Meanwhile, Jim Breen, spokesman for LANL, said that "every indication we have, from all the monitoring, which is year round, is that if there is any pollution of any sort, it is well, well below the standards (of safety) set by the government."[28] Secrets were thus replaced by the contest between diametrically opposed discourses of warring parties: one deeply distressing and accusatory; the other excessively calm and unapologetic. In the end, the effect on the audience remained incomprehension and a sense of hopelessness.

Because of its extensive network of facilities, Los Alamos was hardly the only place in New Mexico with environmental issues. In Albuquerque, fears also centered on the contamination of water as a result of radioactive waste produced by SNL, Kirtland AFB, and UNM. George Tucker, Director of the Health Physics Division of Sandia Corporation in the 1970s, explained that the 2,400 cubic feet of waste that the Lab produced each year was, for the most part, not dangerous and "the radioactivity is high enough that you can't ignore it but it's low enough that it doesn't warrant a major effort."[29] The conclusions of journalist V. B. Price more than thirty years later, however, differed radically from Tucker's. According to him, military dumping of toxic and nuclear waste into the city's water supply since World War II had contaminated a substantial part of southern Albuquerque, including the heavily populated South Valley, which he calls "the city's major groundwater sacrifice zone."[30] The neighborhood is one of the fifteen New Mexican sites on the National Priorities List (NPL)—the most hazardous sites across the United States also known as superfund sites. The EPA called it "the top priority site in New Mexico."[31]

The reason for this contamination is Sandia and Kirtland's proximity to the Tijeras Arroyo, which runs through the base before emptying into the Rio Grande and is "a major recharge area for the aquifer that, until 2008, supplied all of Albuquerque's drinking water."[32] The arroyo is believed to have been an illegal dumping site for years. It was lined with cement after a scandal of blue baby syndrome in Mountainview in 1981.[33] The area just north of this neighborhood is mostly wasteland, junkyard, and small industries, but it is also residential—the east San José neighborhood and the predominantly African American neighborhood called John Marshall. In 2002, the New Mexico Environment Department (NMED) levied a determination of an imminent and substantial endangerment to health and the environment against SNL. Yet, as Price points out, one of the problems with polluting corporations is that the threat of being caught and having to pay a fine and pay for cleanup is not a strong enough deterrent since these costs often seem a trifle compared to the billions made while polluting.[34] As long as the costs do not exceed the benefits, production and growth continue.

Outside of Los Alamos and Albuquerque, other areas (not all mentioned here) inherited a share of the weapons complex's environmental legacy. In Socorro, more than forty tons of depleted uranium shells were tested at a firing range behind the New Mexico Institute of Mining and Technology by TERA from the 1970s to the early 1990s. Residents are concerned about the depleted uranium used in munitions, which is a highly toxic substance. In Carlsbad, the Gnome test left more than forty thousand tons of contaminated salt, earth, equipment, and buildings. The AEC buried it all in the cavity and covered it with soil in 1968. In 1972, when inspectors revisited the site, the waste was uncovered and exposed so DOE decontaminated the site once again, burying salt and dirt 1,100 feet underground. In Farmington, at the Gasbuggy site, barrels of contaminated water and sludge were also buried in the crater in 1978. Three thousand pounds of waste were sent to the Nevada burial site in Beatty while radioactive equipment was steam cleaned before being put back into use. However, tests showed that some tritium remained in the soil there. On Chupadera Mesa, the land was still contaminated with residues from Trinity over three decades after the blast. The *Albuquerque Tribune* attested in March 1981 that farmers and ranchers north of WSMR raised their crops and cattle on land containing up to eighty-six times the normal background level of plutonium. Bob Ramsey of the NRC declared, "Frankly, we don't know how to approach decontamination of that large a piece of land. We don't have a specific plan."[35] It is no wonder residents have expressed increasing concern over the incidence of cancer in the area.

In Grants' uranium belt, the remnants of the uranium industry endanger the health of all who live in the vicinity of the forsaken mines, mills, and tailing piles. The thirty-million-ton mountain of waste at the Kerr-McGee mill, the largest tailings pile in the United States, has been described as a manmade mesa.[36] These remains pose an immense environmental threat as mill tailings retain 85 percent of the original radioactivity in uranium and some of them lie close to water sources. Kerr-McGee's site, for instance, lies sixty feet from the San Juan River, the major water source for the local

Navajo population.[37] Tailings were also used as building materials for roadways, tribal buildings, and playgrounds. Near Shiprock, a retired Navajo miner lost two grandchildren to birth defects before learning that his house was built on radioactive waste.[38] In 1978 the Uranium Mill Tailings Radiation Control Act classified mills that were eligible for federal cleanup—called Title 1 plants—as those that had sold uranium exclusively to the AEC prior to 1971. Only one mill, the Phillips mill at Ambrosia Lake, which had closed before the commercial market period, qualified for federal cleanup.

In addition to the dangerous refuse uninterruptedly produced by the industry, another threat was highlighted by the Chernobyl disaster in 1986 and by the Fukushima catastrophe in 2011: the risk of a nuclear accident. The probability of a major accident at LANL, SNL, WIPP—which did occur in February 2014—or WSMR is a constant threat to local populations. Numerous unpredictable natural causes figure among the most worrisome scenarios. In 2000, for instance, the ravaging Cerro Grande fire uncovered three hundred toxic sites in Los Alamos.[39] In Albuquerque, flash floods could carry ground and ground-water contamination to neighboring communities. More troubling still is that when an environmental catastrophe occurs, there is a chance for it to go almost unnoticed, as was the case in the spectacular Chuck Rock spill of July 16, 1979. An earthen dam collapsed at the United Nuclear Corporation Church Rock uranium mill, releasing one thousand tons of radioactive mill tailing and 93 million gallons of acidic and radioactive wastewater into a creek that flowed into the Rio Puerco. The waste flowed downstream for at least eighty miles, past the homes of some 1,700 Navajo people. This was the largest radioactive accident in the nation's history according to the US Geological Survey, but most Americans never heard about it because it affected people with little political agency. Researcher in nuclear and environmental justice history Linda M. Richards writes:

> Earlier that year the nuclear accident at Three Mile Island power plant made national and international news; the Church Rock accident did not, even though it distributed more than three times (46 curies) the radiation levels of the Three Mile Island accident (13 curies). For Church Rock residents there was no state of emergency, no evacuation, and limited alternative water supplies.[40]

Church Rock Chapter Vice President Robinson Kelly testified "that his uncle died of 'cancer of the foot' [years later] which he believes was the result of wading through the acidic, radioactive effluent in the Puerco to gather up the family's sheep." Many of the local residents who entered the water that day later developed blisters and sores on their feet and legs like Kelly's uncle.[41] Clean-up operations took place in 1980 and 1981: 3,500 barrels of waste material were retrieved but this is estimated at only 1 percent of the waste and very little spilled liquid was pumped out of the water supply according to Paul Robinson, director at the Southwest Research and Information Center.[42] This incident and its lack of media coverage in the same year the Three Mile Island accident made the headlines establishes the connection between environmental issues and neocolonialism, for how else can the discrepancy in the treatment of the two events be accounted for?

Environmental Justice and Nuclear Neocolonialism

The fact alone that the nuclear weapons complex is so developed in a state with a demographic composition such as New Mexico's, in close proximity to Native American reservations, ruins, shrines, and sacred lands, or to centuries-old Hispano communities can be identified as a factor of environmental injustice.[43] Nuclear activities have long been determined as among the most dangerous and riskiest of enterprises, so the choice of New Mexico as the place to extend the country's weapons program, carry out tests, leave dangerous remains from uranium mining, and build its first, and unique, permanent underground radioactive waste repository can hardly be interpreted as anything other than a form of internal colonialism. New Mexico is the only state that has supported the entire cradle-to-grave nuclear economy from mining to storage, and yet, it remains one of the poorest states in the country. Native Americans and Hispanos represent more than 10 percent and 47 percent of the state's population respectively.[44] The two groups make up the statistical majority in a so-called minority-majority state, but in terms of power and economy, they are minorities.[45] Throughout the state, nuclear neocolonialism is manifested in the interactions among the nuclear industry, the often relatively poor, non-Anglo communities, and environmental issues.

Other authors have reached the same conclusion. V. B. Price explains that seeing New Mexico "as a national sacrifice zone helps us to understand environmental racism." His alarming study of the state's environment lays open the damage caused by the scientific conquest that he emphatically calls a "nuclear colonization" targeting "marginalized populations, the urban and rural poor of all races and cultures," pointing out that "poor people, lacking economic clout, are relatively voiceless and hence less troublesome to corporations and governments than better-heeled Americans are." Pollution in the South Valley is a telling example as it has long been a poor area with a majority of Hispano residents, 70 percent in the 1980s, 50 percent of whom lived below the poverty line.[46] Price designates poverty as the main characteristic of the affected populations over race. Oftentimes, however, the two combine and reinforce the notion of injustice, anchoring the question in the nation's core struggles of identity, responsibility, and relations to minorities.

LANL is surrounded by Hispano communities in Española and the Chimayó Valley. Yet, of the interviewees of the Impact Los Alamos Project, none addressed LANL's environmental impact beyond saying that they believed the Lab was in control or that they had only heard rumors. Paul Frésquez remarks that he thought "a lot of people dump their trash in the Rio Grande."[47] Ramón Frésquez did not believe the Laboratories had anything to do with the high number of tumors and cancers, insisting that "Los Alamos is a good neighbor."[48] Interviewees were distinctly more vocal on economic than environmental concerns because of how much they had come to rely on the Hill. Furthermore, worrying about the environment may be construed as the luxury of those whose economic stability is ensured. A former member of the Santa Fe City Council summed up the conundrum in the following terms:

> You downgrade Los Alamos because of its agenda, but Los Alamos
> has provided a terrific base for the people of northern New Mexico.
> Yes, we didn't go to Harvard and a lot of people didn't graduate high
> school at Española High, but they make a heck of good living at Los
> Alamos. They have beautiful homes in Chimayó, everywhere, built
> on what they have made at Los Alamos. That's not to say there is not
> pollution, and that they don't need to tighten up what is going on up
> there but to condemn the whole thing . . . they need our support to
> clean up and keep it clean.[49]

Prioritizing economy over ecology and having faith in science and technology to find viable solutions are widespread opinions. Talking about nuclear waste, interviewee C. L. Hunter assures, "We will contain these problems."[50] Others address the capital importance of knowledge, whether they profess a lack thereof so as not to venture an opinion on the subject, or attest that their privileged access to information allows them to say, "The Lab has done its part in trying to provide a safe environment for everyone." Leroy Martínez professes that "working at Los Alamos opened [his] eyes on what is going on in research and development, what gets funded, what direction the US is going as far as weapons go, and environmental concerns. . . . If there were more hazards before, it was because of a lack of knowledge."[51]

Relating to environmental issues is usually more complex than relating to economic ones because the effects are not immediate. Yet, one must not ignore the relationship many New Mexicans have to their homeland. Some former employees or their children became activists. Their concerns crystallized on economic and cultural matters, which often encompassed environmental elements. Writer Juan Estevan Arellano, whose father was employed at Los Alamos in the 1940s, became an advocate of traditional agriculture and *acequia* (communal irrigation systems) culture. He refers to Los Alamos as "an aberration."[52] Grassroots organizations, such as Juntos: Our Air, Our Water based in Albuquerque, aspire to represent Latino families in their call for clean air and water. In September 2011, a poll by the New Mexico–based Latino Sustainability Institute revealed that more than 90 percent of Hispano voters in New Mexico were "worried about water scarcity and "pollution of drinking water, lakes and rivers."[53] A large majority also supported additional public lands designated as national monuments, a way to preserve the area and create employment for the local population.

Environmental issues are fundamentally linked to the socioeconomic status of New Mexico Hispanos and to the future of their traditions. Joseph Masco, who uses the term *nuevomexicano* in his book *The Nuclear Borderlands*, cites one Hispano who wove together the threads of environmental hazards and social justice in the following testimony:

> Who knows what is working its way down that mountain, into our water supply, into our soil. . . . In the early days they weren't too careful up there. Today, a lot of men from the Valley work up in

> Los Alamos as plumbers and electricians, they're afraid of what they
> might dig into when they're working underneath people's houses, or
> what might be in the soil. We've always done the shit work up there
> and we don't know what the effects are.[54]

Contrary to Price, Masco addresses environmental justice through the
prism of race rather than poverty, arguing that the Manhattan Project created "an enormous biosocial experiment" in northern New Mexico. Some
New Mexicans have wondered if they were being treated as guinea pigs like
other populations during the Cold War, for example, the *Hibakushas* (surviving victims of the 1945 atomic bombings of Hiroshima and Nagasaki),
downwinders, atomic soldiers, Pacific islanders, or plutonium-injected
patients.

Historian Valerie Kuletz, for her part, uses the terms *environmental
racism* and *nuclearism*, which she also defines as a form of "internal colonialism."[55] Colonialism and atomic ventures were more explicitly connected
in other contexts such as that of the French atomic tests conducted in Alge-
ria (1960–1966), a major colony in the French empire at the time. Yet, the
comparison with the United States is not without relevance if one considers
that the original inhabitants of the region feature among the most affected
populations. Kuletz writes about how Native American communities and
the actors of the nuclear weapons complex interact in a land that the former considers as "a crucial link to cultural survival," while the latter have
made it into "a landscape of national sacrifice." She contends Natives could
at best be hired on "low-paid jobs to help build, maintain, and clean the
emerging cities" or at worst "were ignored completely—rendered invisible
by a mixture of racism and a perception of desert lands as vast, uninhabit-
able wastelands."[56]

New Mexico is a prototypical illustration of her stance and the Church
Rock spill even more so; the Navajo families who were affected were rendered invisible by their absence in the mainstream media. As antinuclear
activism gained momentum in the state, activists developed their own
vocabulary, also borrowing from the colonial lexical field. Native American
activist Winona LaDuke began using the terms *radioactive colonialism* in
the 1980s to describe the dumping of radioactive waste on Indian reservations.[57] Activist groups and nongovernmental organizations (NGO),
such as Concerned Citizens for Nuclear Safety, Southwest Research and

Information Center, and Nuclear Watch New Mexico, now provide much information on nuclear pollution in New Mexico with sophisticated expertise. Using the same words as scholars, the Nukewatch New Mexico website refers to the state as the "ultimate national sacrifice zone."[58]

One symbol of the industry's relationship to Native cultures is the adjacency between contaminated areas and sacred grounds, ruins, or shrines. Los Alamos lies between Bandelier National Monument, which is regarded by modern Cochití people as their homeland, and the Puye ruins on Santa Clara Pueblo lands, where the ancestors of the pueblo people carved dwellings in the cliffs. In between, the land occupied by the Laboratories is regarded by the San Ildefonso Pueblo people as their homeland. Probably one of the sites that best epitomizes the clash between the culture of nuclearism and that of Native peoples is Tsirege Pueblo, home of the ancestors of the modern Pueblo Indians, the Anasazi, which lies on LANL property next to Mesita del Buey. The mesa, also called Material Disposal Area G, was the largest disposal site on the Hill for wastes contaminated by radionuclide, toxic or explosive chemicals, and classified materials buried in pits or shafts dug into the ground.[59] When Area G began operations in 1957, five Native American ruins were destroyed. The replacement of archeological items with radioactive waste is reminiscent of the swapping out at Santa Fe's Museum of Anthropology of Native exhibits for pictures of the atomic bomb in Hiroshima at the end of the war, materializing the semiotic transition from ancestral knowledge to modern destructive science.

By the same token, Bayo Canyon, west of San Ildefonso was used from 1944 to 1962 for a series of 254 tests called the RaLa (for radiolanthanum) program to improve the implosion design. Fallout was documented as far as seventy miles downwind and measured by the US Air Force over San Ildefonso, Santa Clara, and Española. In later years, some tests were allowed only when the wind would blow away from the town of Los Alamos.[60] To make matters more disturbing, firefighters from Zia and Jémez pueblos, provided with radiation badges, gloves, and burlap bags, were hired in 1963 to clean up the site at the end of the experiments. They picked up the debris and removed truckloads of refuse to Area G.[61] The first public mention of the program appeared in a LASL newsletter describing the cleanup operations, but the first concerted efforts to inform the Pueblo occurred in 1994 when LANL reviewed the program.[62] Thus, not only was fallout from the RaLa tests knowingly likely to affect the Pueblo

Indians inordinately, but cleanup operations were presented as a profitable job opportunity for the Native communities, making Bayo Canyon a striking illustration of environmental injustice.

Authorities at the pueblos have long been concerned about contamination from the Lab's activities. In 1979, one San Ildefonso spokesman said, "They tell you there's no danger, but I know better. There's radiation dumps all over the place and rain puts radioactivity into the soil, that solid rock thing is bull—the tuff is porous as hell."[63] In 2008, plutonium was found in wood ash from an interior woodstove at San Ildefonso, and high doses of strontium-90 were found in dust samples from Picuris Pueblo. LANL scientists concluded that the radioactivity "did not come from the Labs, but rather from nuclear fallout during worldwide atmospheric nuclear testing in the 1950s or from naturally occurring radiation levels in northern New Mexico," which are generally high.[64] A Pueblo member, Darryl Martínez, explains how the community now has its own environmental program to monitor the water. Their prime concern, however, is the Labs' expansion. Darryl wonders whether they would have to settle elsewhere like his ancestors had been forced to do when the weather irreversibly changed or once the fields were overgrazed. He believes LANL could be a new reason for departure, especially considering that a worrying number of people in the community have died of cancer.[65]

Suspicion has grown to such an extent that "many Pueblo do not trust the Indian Health Service to record accurately causes of death, claiming that some people who have died of cancer have had other causes of death listed on their death certificates, which has skewed the official cancer rates for those living in the shadow of Los Alamos."[66] One Pueblo member testified that "in the old days, LANL workers were given cigarettes at pay day, sort of like a bonus. Now, the Lab blames the cancer on the smoking." Elmer Torres, a tribal liaison between the Pueblo Indians and LANL and a former governor of San Ildefonso, asserted that LANL's plans were generally shared with all the people. But when researcher Annie G. Ross showed the document to Pueblo people, they had never seen it and could not read it because the convoluted language that was used. Myron Gonzales testified that "the tribal councils have gotten used to the money, and do not want to ruffle the feathers of the Lab," showing how politics and economics are interlocked. Governors were allegedly paid to keep their job, keep quiet, and keep the others quiet.[67] Like Hispanos who depend on the nuclear industry for jobs, Native communities struggle with its costs and benefits. One member of the San Ildefonso tribal council avowed his people were "concerned about radioactive pollution from Los Alamos, but can do little because they are economically dependent on the Lab."[68]

In the Pueblo of Ohkay Owingeh (formerly known as San Juan Pueblo), Naomi Archuleta, program manager in the Office of Environmental Affairs, expresses her community's concern over nuclides in the water, the air, and the soil, and especially in the adobe they use to build their homes. In 2000, one concern was the ashes from the Cerro Grande fire that would land on the Pueblo. The most striking aspect in Archuleta's discourse is how she repeatedly asserts that her community would never move, that they had been here forever, and would be here forever, contrary to what Darryl Martínez said. That is the fundamental difference between Los Alamos and its surrounding communities, Archuleta emphasizes,

> This land is not their home. They don't care about what they leave behind them. They can pack up their house in their cars and go back wherever they came from not caring about what they leave behind. They don't care, but we do. We will be here for many more generations. I don't see any tribe ever moving anywhere else.[69]

Another Pueblo member concurs, saying, "We are not mobile to be refreshed elsewhere. It is not like we can move to our second home in some other part of the country. This is home. This is our place."[70] Such statements prove that the Pueblo's sense of rooting has not died out but has been reinforced with a commitment to environment protection. In 1993, at a meeting of the Southwest Indigenous Uranium Forum, Manuel Piño from Acoma Pueblo reported that a uranium company had once asked the Paguate community if they would be willing to move because the village was thought to be built on the richest uranium deposit in the area. His reaction was to say, "This village is sacred land. You don't move a village like that for economic incentives."[71] These attitudes are reminders of Native peoples' centuries-old history of resistance to uprooting and of the ancestral cultures of the land, which are still strong identity markers in New Mexico, even though economic pressures to lease or to sell the land and to exploit its resources have grown.

Uranium, the Epitome of Nuclear Racism

The epitome of environmental and racial injustice undoubtedly is uranium mining, which has even been classified as a purposely perpetrated technological or human-caused disaster. Several publications have admirably addressed the story of Navajo uranium miners; nonetheless, one cannot write about environmental justice in New Mexico without addressing this issue.[72] The Navajo Nation, the largest reservation in the country, has an estimated one thousand abandoned uranium mines and four former uranium mills on its lands, which contributed 13 million tons of uranium ore between 1945 and 1988. The largest uranium mine in the world sits next to Mount Taylor, also known as *Tsoodzil* (the Turquoise Mountain), one of the four sacred mountains for the Navajo Nation and a pilgrimage site for Pueblo people.

After uranium was discovered so close to the reservation, young men gratefully embraced new job opportunities: blasters, timber men, "muckers," transporters, and millers were paid minimum wage or less. For many, this was their first contact with the wage economy. Former miner Terry Light recalls, "The company came around and said there were mining jobs opening up, but they didn't tell us a thing about the dangers of uranium mining. . . . The white men really took advantage of the Navajos who needed jobs."[73] Years later, workers started falling sick with lung diseases, such as silicosis, tuberculosis, pneumonia, emphysema, and various cancers. In 1952, William Bale and John Harley's work on the concentrated effects of the energy released by radon daughter isotopes led to the definition of *working level* (WL). For years, though, scientists were forbidden to publicize concerns about the health hazards of radon in uranium mines. After reports showed that ventilation was necessary, the state gradually enforced regulations at the end of the decade.

The Kerr-McGee Company hired about seventy Navajo miners and thirteen Anglo supervisors. Although the miners' exposure was within the national standards per hour and per week, their yearly exposure exceeded safety standards.[74] George Tutt, who started working in the mines in 1949 at age fifteen as a "hand mucker," shoveled uranium waste and ore by hand without mask or gloves for $2.50 per hour in 1960. He began experiencing respiratory difficulties a few years later. Because the Bureau of Indian Affairs prohibited unions on the reservation, Navajo miners earned $0.75

per hour less than union uranium miners in 1963 and the difference was more than $1.00 in 1968.[75] Most miners explained that they did not know anything about safety. Floyd Frank even wondered if they were "disposable to the government." Tommy James, miner in the 1950s in Shiprock, describes how the smoke could give headaches and nosebleeds. He depicts his job as "slave work" and underscores the differences between Navajo workers, who had to build their own tents and for whom the companies "never prepared anything," and Anglo surveyors, mechanics, and office workers, who had cabins and showers. He also describes how waste was spilled out of the mill and flowed into the river. Joe Ray Harvey, who started working for Kerr McGee in 1961, complained about the lack of ventilation. Spouses and children were exposed as well because they washed the clothes and sometimes lived onsite in the tents. The children played in the radioactive dust and drank contaminated water. Minnie Tsosie, widow of a Navajo miner, reveals that both her daughters had to have their uteruses removed.[76] One relative of a deceased miner lamented in 1979, "Where the uranium slug has been dumped children wade in the water but break out in sores."[77]

In December 1978, ninety-two Navajos and one Acoma filed a lawsuit against six branches of government[78] to put an end to uranium production. The plaintiffs contended that many projects in the past decade (leases, mining, and reclamation plans) had been federally approved without environment impact statements, which was a violation of the National Environment Policy Act of 1969. One of the plaintiffs, Mary C. Largo, a seventy-eight-year-old Navajo woman of the Dalton Pass Chapter—an area under lease northeast of Gallup—signed up after drilling began on her land allotment without her permission: "I never put my thumbprint to anything. All at once the trucks and drills started coming onto my land, but nobody from the company told me anything about what they were going to do." Twenty uranium exploration holes surrounded her traditional hogan, several of them less than one hundred feet from her front door.

According to DOE estimations at the time, "persons residing within a half-mile radius of uranium mill tailings piles have a 100 percent greater risk of contracting cancer than the general population." The local BIA representative told Largo that the mining operation would bring her money but that she should not let her sheep feed on the grass or drink the water contaminated in the uranium exploration. By 1979, twenty-five Navajo uranium miners had died and an additional twenty-five suffered from radiation-induced lung cancer.[79] In April that year, thousands of protesters, not only Navajo and Pueblo but also Hispano and Anglo, gathered at Mount Taylor to demonstrate against the exploitation of uranium deposits on Native lands. As a result of their protest, uranium miners were included in the 1990 Radiation Exposure Compensation Act (RECA) that provided monetary compensation to atomic veterans and downwinders. By spring 1994, 155 Navajo workers or families had been compensated. However, the Act did not cover all the diseases from which the workers suffered; it was difficult for widows to have their traditional Navajo marriages recognized by the Department of Justice; and the millers who were exposed to uranium, silica dust, and concentrated uranium oxide were excluded from compensation.

As illustrated by the plight of uranium miners, the environmental impact of the nuclear weapons complex went far beyond affecting wildlife in New Mexico. This legacy of scientific conquest also is connected to other issues, such as ethnicity and poverty, two key components of colonialism. Radioactivity is pervasive. It contaminates all, yet all are not exposed equally. These differences are evidenced through the geography of contamination that presents a correlation between "hot" areas, sanitary problems, and communities largely inhabited by minorities plagued by economic difficulties. The South Valley and the East Heights in Albuquerque or the Hill and its radioactive particles traveling down to the Valley are two of the most evident manifestations of this phenomenon. The first environment-related scandals and lawsuits shed light on neocolonial practices to superficially treat radioactive waste or clean up radioactive sites. From the start, Native Americans were particularly active in voicing their protest because their cultural practices put them at greater risk. Yet, protest among the Hispano population was brewing as well. These movements focused primarily on economic discrimination as, for many New Mexican workers at defense-related facilities, it was at the workplace that they were exposed the most.

Ten

The Sociocultural Impacts
of a Scientific Conquest

The power elite theory of sociologist Charles Wright Mills designates a triumvirate of powers that dominate US society: the corporate, the political, and the military.[1] In the 1950s, the increase in militarism was profitable to each power as it enabled the government to put forth a strong vision, the military to thrive, and corporations to make bigger profits. Caught in this mechanism were those Mills calls "the masses." This triumvirate was the foundation of the nuclear weapons complex defined by Peter Hales as a "powerful consortium of institutions ranging across the worlds of business, government, and the military, devoted to self-perpetuation and eventual colonization of the American democracy."[2] The theory thus mirrors the pillars of the scientific conquest: the nuclear golden goose, the federal sponsor, and secrecy imposed by national security. By putting their faith and fate into the hands of corporate and government interests, New Mexicans looking for jobs from the largest employers were integrated into the "masses." The mining corporations, SNL, LANL, the government, and the military nourished the mechanisms of economic growth and crisis to maintain control over a social structure relying on this labor force. At the same time, the state lacked a solid economic development strategy to compensate for growing inequalities. As the ups and downs in the nuclear industry drew peaks and valleys in the region's economic line graph, local workers were the first to suffer the consequences.

Who Benefited from Scientific Conquest?

After the first boom stage in the late 1940s and 1950s, New Mexico's economic growth slowed down in the 1960s because the nation's industrial expansion had little impact on states with small manufacturing sectors and defense money was redirected to ground war. Then, during the 1970s, federal expenditures and employment soared, with a sharp increase in DOE and DOD spending. This period of exceptional prosperity was coupled with the second uranium boom fueled by the US crusade for energy independence and skyrocketing energy prices due to the OPEC oil embargo. In the early 1980s, however, "the robust economic expansion of the 1970s came to a screeching halt with the first break in oil prices . . . and troubles in the US nuclear power industry triggered by the Three Mile Island incident."[3] The uranium industry suffered its second bust period. Mining employment was cut in half between 1981 and 1987, and nonagricultural employment fell for the first time in twenty-one years in 1982. Despite these setbacks, New Mexico's economic growth throughout the 1980s still exceeded the performance of the US economy.

According to historian Michael Welsh, "New Mexico had outperformed the nation as a whole in both population and income growth, . . . the state should be a case study for both economists and historians interested in the ability of the American economy to deliver on its promises."[4] Yet, the positive spin on these figures hinges on two key elements: whose perspective and what promises are taken into consideration. The repercussions of variations in the government and extraction sectors on employment consequently impacted poverty rates, which dropped to 17.5 percent between 1969 and 1979 but increased to 38 percent in the 1990s.[5] However, improvements in poverty rates could also be partly accounted for by the incoming waves of highly educated elites who came to work at the Labs and other research centers and helped boost the median income. Lower poverty rates do not necessarily equate less poverty for the poor but rather a larger proportion of more affluent individuals in the state.

There is no denying the positive impact of such employers as the LANL, Sandia, and the DOE.[6] For fiscal year 1990, LANL and SNL alone had a statewide economic impact of $3.67 billion and $6.65 billion respectively and provided close to twenty thousand jobs.[7] Yet, notwithstanding the spin-off effect on local businesses, the work opportunities for locals, and the beneficial effect of immigration on entrepreneurship and the local consumer market, after decades of exceptional growth exceeding national performances, New Mexico is still, in the twenty-first century, one of the poorest states in the United States.[8] So, the question arises: who were the beneficiaries of its scientific conquest?

With each step in the scientific conquest, new branches of the industry sprang up, and newcomers moved en masse to Los Alamos, Albuquerque, Alamogordo, Grants, and Carlsbad. The influx of domestic migrants drastically underscored existing inequalities and created new divisions between elites and poorer classes. In 1997, a report aiming to define poverty in New Mexico observed that, although there had been variations in the state's poverty rates, New Mexico had consistently ranked among the nation's poorest states since statehood. Poverty can be attributed to various origins and responsible parties (local political leaders and their strategies, for instance), so the nuclear industry, narrowly understood, does not merit all the blame for all the difficulties that New Mexico's impoverished populations face. Rather, the state's evolution in the late twentieth century demonstrates how inequalities grew owing to the unguided, inequitable distribution of returns from the weapons complex. The 1997 report found that the benefits of economic growth did not "appear to automatically transfer to those in poverty" because "educational and training opportunities must be provided for local workers to reap the benefits of a growing economy."[9] In his thesis in 1989, historian Chris Dietz explains the biases behind the indicators of economic progress:

> When jobs and money generated are used as the prime indices for evaluating a science facility's impact on a region, a kind of tautological pretension occurs: jobs and money are needed in the region; the new facility begins operation; jobs and money are generated; therefore, the impact is beneficial. These kinds of indices are expedient, easily available. But they are also simplistic, refining complex social and cultural relations to facile singularities. . . . Native New Mexicans have primarily participated at LANL in unskilled and skilled labor positions. Few Native Americans or Hispanos work in top level positions.[10]

To include such subtleties in their analysis of the determinants of poverty status, the authors of the 1997 report separated the New Mexico–born, other US states–born, and foreign-born populations. Statistics revealed that New Mexico residents born in other states, "regardless of their educational attainment, have a predicted probability of being poor that is lower than that of college educated local-born New Mexico residents." This shows that the pattern of chronic poverty in the state is due, in part, to the competition with outsiders that even education cannot remedy. In general, education pays off as the number of people reporting incomes below the poverty threshold decreases with the level of education attainment, but the correlation between the two is not "straightforward": it is "mediated by race [and] place of birth." The median wage income for New Mexico–born residents was around $16,000, while it was $22,000 for US-born domestic migrants. Even among individuals without any education, the difference was about $2,000 ($4,000 for college graduates). These figures suggest a clear pattern of separation between the local-born poor, who seem to struggle to get out of poverty even through education, and newcomers from other states, who seem not only to be protected from poverty in most cases but also to have higher income with the same educational attainment. Without synergy among employers, surrounding communities, and local political leaders, the positive effects of the labs' presence are bound to be undermined by the vicious effects of discrimination.

Speaking English was also identified as an important factor because foreign-born individuals with a good mastery of English were less likely to be in poverty than local-born New Mexicans whose mastery of the language was poor. Yet, again, "residents who were born in other U.S. states have a distinct advantage over everybody else regardless of their ability to speak English"—a 6 percent chance of being poor for English-speaking domestic migrants; twice as much for local-born English speakers.[11] Bilingualism was one of the major difficulties for locals in Los Alamos, both at school and at work. Work routines and protocols could be difficult to understand when group leaders spoke fast or used complex vocabulary to give instructions.[12] Speaking Spanish was discouraged because it was important to parents that their children be proficient in English. As a result, in a family of seven or eight children born in the 1950s and 1960s where the parents worked at Los Alamos, typically, the four younger children would speak

English while the four older children would speak Spanish.[13] Some of these English-speaking children who were born in Los Alamos but had roots in the Spanish-speaking communities of the Valley chose to attend Spanish classes at school to be able to communicate with their grandparents.[14]

Disparities in income can be mapped out in the state between the counties that are part of the nuclear complex and those on the periphery. In parallel, the ups and downs in the whole industry had correlative repercussions on poverty rates for New Mexicans who did not have the job security of the out-of-state PhDs who came to the state after being offered lucrative positions in the industry. While having a positive impact on the state's economy on the whole, the hiring and wage policies of the Laboratories thus also contributed to increasing inequality. For example, if one looks at the evolution of the number of employees and payroll at SNL over ten years between 1965 and 1975, one can note the decrease in the number of employees (from 7,120 to 5,542), while, at the same time, both payroll and plant assets were significantly higher (from $69.50 to $95.30 and from $161.30 to $273.00 respectively), meaning that employees were fewer but better paid.[15]

Lab scientists and technicians earned salaries unheard of in the rest of the state. Even security guards and janitors paid minimum wage or slightly above often considered their situation as immense improvement. Therefore, whatever their work positions, those employees came to be viewed as privileged by others. Danny Martínez considered inequality as a legacy of the wartime era. He states,

> When Los Alamos first began, hundreds of educated gringos were imported into northern New Mexico and the people from the Valley were brought in as janitors and stuff. It set up a big social difference that still exists today. . . . Even a secretary at Los Alamos earns more than a carpenter that has a physical job to do. It is incredible the economic segregation which exists in Chimayó. Farmers who earn $3,000 a year compared to people who make $60,000. You have to feel sorry because you know that this is where you came from: you were in their shoes a few years ago.[16]

Martínez's comments reveal that competition pervaded communities, pitting wage earners against each other in the tense labor market for locals.

His examples are socially, but also culturally, significant. The carpenter and farmer are symbols of the region's past, whereas the labs represent the future, to which only a fortunate few have access.

An Island of Privilege

Nowhere else in the state do inequalities appear as unequivocally as in Los Alamos. The town and county hardly have anything in common with the rest of New Mexico, be it the population, the standard of living, or the specific corporate culture and history. In 1969, journalist Walter Briggs underlined the "stark contrast between the ancient and the futuristic," establishing a dichotomy that opposes "some of the most modern architecture" to the "panorama of some of the oldest of Indian ruins" and the "600 doctors of philosophy study[ing] scientific problems not even envisioned a dozen or so years ago" to the "Indian and Spanish-American communities living almost as their ancestors did."[17] Juan Esteban Arellano stresses the same disconnection, comparing Los Alamos to the floating island in *Gulliver's Travels*.[18] The island metaphor was also chosen by Chris Dietz, who writes about "an elite, Anglo island in New Mexico," "a place of great privilege and a place of evil." One of the respondents in his study commented that people expected a special place of "evolved human potential" where people "should have all the answers." When it becomes clear that Los Alamos is not all they expected, they "feel cheated." The island image is thus coupled with the idea of a mirage, mirroring Arellano's reference to *Gulliver's Travels*.[19]

This was reinforced by the socioeconomic contrast, as 90 percent of the population in Los Alamos County was Anglo in 1969 and only 2 percent were below the poverty line compared to 38 percent in neighboring Sandoval County.[20] In recent years, the divide has not been reduced; on the contrary, Los Alamos County placed third richest county in the nation with a median household income of $110,000 in 2011. That same year, with a 21.5 percent poverty rate, New Mexico was the second poorest state in the country.[21] On maps showing poverty rates in the state, Los Alamos stands out as a speck of white in a pool of darker colors.[22] Historian Jo Ann Shroyer calls the town an "anomaly": "a predominantly white society" and "affluent community" in a state that is multicultural and poor. She points out that Santa Fe might be just thirty-five miles away, but it is "at least a light-year away in terms of attitude and focus" and "alive with antinuclear activists who keep a suspicious watch over their neighbors on the mesa." One of her interviewees at LANL likened the two towns to Sparta and Athens, two cities at war with each other but also fundamentally needing each other to survive.[23]

In 1965, the Public Works and Economic Development Act encouraged communities to "work in partnership with neighboring communities, the States, Indian tribes, the private sector and the Federal Government" to implement "comprehensive economic development strategies to alleviate economic distress."[24] The plan was based on the observation that the five counties around Los Alamos, except Santa Fe, suffered from high unemployment rates and low median family income, designating them as economically depressed.[25] These efforts were not deemed sufficient by certain observers, such as Harold Gibson, who told Siegfried S. Hecker, LANL director from 1986 to 1997, at a town meeting, "For fifty years you've been talking of community outreach, but you haven't done it here, all you have done is hire people, fire people, and you have been a brain drain rather than create a brain trust." According to him, the Valley suffered because the Lab did not see it as a viable partner.[26]

Many argue that Los Alamos has been segmented by education and occupation rather than people's origins. Katrina Mason points out that "academic success was the leveler" and that "scholastic achievement, not family background, determined one's place in the children's social strata."[27] This statement is valid when one considers the community from the inside, but, for those on the outskirts, the stratification along geographic and ethnic lines was a reality. Education and social backgrounds were connected. Secundino Sandoval, son of a craftsman from Albuquerque, explains that if "your father was a carpenter, you were in a lower echelon . . . , which caused him a lot of fights at school." He subsequently mentions the stereotypes outsiders who came to work in construction had about New Mexicans, saying, "In Texas or Oklahoma, back in those days, if you were of a different race, dark skin, dark hair, you were Mexican. They categorized everyone under the name Mexican, whereas my father always said, 'Hey, we're Spanish.' Spanish-American, that was the term used then."[28] One can note that the designation "Mexican" is perceived virtually as an insult. Many New Mexicans had internalized the social hierarchy that presents Anglos at the top, Hispanos (or Spanish Americans), Native Americans, and last, Mexican immigrants.[29] Ethnic categorization is common in the United States, but this classification reveals profound divisions within New Mexican society, despite its reputation of colorful multiculturalism.

In a 1949 *National Geographic*, under a picture showing a classroom, the caption read, "All Los Alamos children seem to know the answer. Aren't their dads Ph.D.s?"[30] Education was, in fact, the main factor of social reproduction, as many children went into physics because it was all they knew from their parents. Nella Fermi Weiner acknowledges that being the daughter of a physicist increased her chances of success. She notes, "We [the kids of the physicists] were not necessarily more intelligent, but we certainly had more opportunities than these other kids."[31] They had access to one of the best school systems in the nation, and, after going off to the best universities, many returned to be hired at the Lab on the best jobs, thus perpetuating the cycle of hereditary wealth. Senni Gallegos of Chimayó, a secretary in the weapons division, confesses, "Sometimes I envy people I work with who have Ph.D.s. They are physicists, and they are always pushing their children to get an education. They have the money, and they can afford to send them."[32]

Social reproduction and prejudice maintained a pattern in which employers turned automatically to other states to look for expertise. Dietz argues that the main reason why Los Alamos had so much trouble "fitting in" was that the Laboratory did not acknowledge any potential of local minority students who would have liked to pursue a career in science. They believed themselves and other states to be the only possible reservoir of talent, discarding local institutions. Santiago Bustamente comments, "The people who do the hiring [at the lab] don't think much of New Mexico schools." He was hired as a draftsman with three years of experience in 1951 but didn't make staff member until years later because people without degrees, he was told, could not be promoted. Yet, there were young Anglos "fresh out of high school who were made staff members right away."[33] Another Hispano employee with twenty-four years of experience at LANL confirms, "Oftentimes, a local contractor would seek some specialized technology out of state because he did not realize the skills were locally available." On a tour of northern New Mexico schools, he asked students whether they wanted to "sweep floors at LANL" or "be on top, a scientist," thus using the mystique of Los Alamos that associates science with success and menial maintenance work with failure.[34]

The politics of the school system reflect the rift between the Hill and the Valley. Paul Emilio Frésquez recalls that when he was the senior student body president in Española, they were "the laughing stock of the state"; his school "had to beg for lightbulbs."[35] In 2006, Española Mayor Richard Lucero threatened a suit over the Los Alamos school district receiving millions of dollars annually from the DOE, while surrounding school districts received nothing, even though half or more of the LANL workforce resided outside Los Alamos. According to the Española school superintendent, the $8 million subsidy enabled Los Alamos to invest nearly twice as much per student. Art Blea, Pojoaque school superintendent, lamented that "his best educators were being siphoned away to Los Alamos by the higher salaries." LANL officials proclaimed that the Los Alamos Foundation, which had been created to address that concern, "celebrated the distribution of $1.2 million to northern New Mexico schools/students in Taos, Santa Fe, Española, and Los Alamos." But the foundation had more than a $40 million endowment balance at the time.[36]

The attitudes of Los Alamosans toward the outside were often shaped by their unawareness or misconceptions of the socioeconomic reality of the region. Maxine Beckman from the Hill talks about going down to the Valley to shop and how people think that, because you are from Los Alamos, "you think you're better or something like that." She couldn't imagine that people did not realize "how wonderful Los Alamos has been to northern New Mexico; it would be a poverty area without it." Her comment reflects a distorted vision of the area, which remained a poverty area despite the influx of money. Similarly, based on her personal experience, she considers the lab to be "a very fair place for everyone," where "if you really want to get a job there, if you're a native or whatever, you've been able to." So-called natives have described experiences that show otherwise.[37] Thus, on both sides, unclear or false impressions shaped each group's perception of the other.

Stereotyping began at the workplace where the type of occupations and the associated salaries could be crossed-checked with the employees' origins. Within the Lab's hierarchy, the two main groups were staff members (SM), scientists with degrees who work on the developmental phase of projects, and technicians (TEC) who do "hands-on" work under the supervision of staff or supervisory technicians. Anthropologist Mary Meyer studied their perception of each other: the former are seen "as being arrogant, 'prima donnas,' impractical dreamers, 'spacey' characters, and workaholics who are unreasonable in their demands and lacking in social skills or personal hygiene," while the latter are seen "as being 'clannish,' unmotivated, slow in their work and primarily interested in working their hours, receiving a good salary, being promoted, socializing on the job, and using all their allotted sick leave regardless of illness."[38] There is an element of truth in that Valley workers generally perceived the Lab as nothing more than a generous employer. Senni Gallegos simply says, "Los Alamos provides my bread and butter."[39] In turn, negative stereotypes permeated communities. Biases and competition cemented the relations between Lab employees and New Mexicans. One of Dietz's respondents talks about class consciousness and the fact that most of the high-salaried, out-of-state staff members had never been around rural, ethnic people, whom they found fascinating but "dirty." Valley people were the butt of jokes at LANL; so much so that the Lab had to issue a memo forbidding them.[40]

In *The 3 1/2 Cultures of Española*, a booklet written by students of Española Valley High School, the authors humorously analyze four cultural groups they name "the Native American," the "Moja'o," the "Cholo," and the "Anglo." The stereotypes regarding the Hill and the Valley are addressed through the story of the Los Alamos envoy that came down from the Hill during the World War II to spread rumors about the secret work underway at the Army camp:

> The success of his mission depended on the paradoxical assumption that the local people were unsophisticated and ignorant, but inquisitive, and that those who frequented the cantina were true representatives of valleyites. The stranger proceeded to mingle boisterously, and drinking with the local "borrachos." . . . Many free drinks later his "secret" was extracted from him. Los Alamos scientists, he disclosed, were researching a new awesome weapon, calculated to hasten the end of World War II: the weapon, windshield wipers for adobe submarines! And thus originated Los Alamos' patronizing attitude towards Española.[41]

The students of the Valley and their teacher found in this story the genesis of their complex relationship with their scientist neighbors, which was based on assumptions of intellectual superiority and inferiority. Another of

Dietz's respondents, for instance, spoke of the "genetics" of intelligence that oppose Los Alamos and the area's students.[42]

Through cross-racial stereotypes, Los Alamos came to be seen as a place where residents "suffer from xenophobia, a fear of being inundated by the 'Brown' hordes from Española," like a medieval castle on a hill, the gates recalling a drawbridge, the residents likened to lords of medieval Europe, and the workers to peasants. The authors of the booklet comment, "THEY live atop the 'Hill' . . . , and the unskilled worker and garbage collector . . . live in the Valley. This mentality was visually reinforced by the unofficial policy of hiring janitors and refuse collectors from Española, and recruiting young men, to be trained in skilled or semi-skilled professions from out of state." The medieval image thus emphasizes the us-versus-them dichotomy that reproduces dominant discursive patterns in US social relations. Ideological and political differences also reinforce the opposition as Los Alamos residents are thought to be "generally ultra-right wing" in a traditionally Democrat state.[43] C. L. Hunter talks about a "rift" between "the *nativos*" and the outsiders. "These guys came from out of town, and they had the money, the good-paying jobs and were attracted to local girls, and fights started," he said. Now, "there is still the rivalry" observable at events such as school games opposing "the rich kids, the affluent society versus the meager lower income society of the north."[44] The recurrent binary discourse gives a simplistic view of Valley-Hill relations, yet this was the consequence of systematic separation policies that influenced the social cognition of both groups.

"As a Minority, It Was Never Enough"

The word *minority* appears frequently in the testimonies quoted in this work. It is, intrinsically, a controversial social concept, as it defines a culturally or ethnically distinct group through its subordinancy to the dominant group without any necessary correlation with population numbers. It is all the more relevant in a minority-majority state such as New Mexico. The concept is so widespread that members of those groups have internalized it as part of their identity. As minority members defended their rights, some members of the dominant group began associating the successes of others with their belonging to minority groups. Starr Beckman professes she was envious of one of her friends whose marital name was Sandoval because she thought being a minority offered better chances. She worked with machinists and technicians, among whom she guessed Hispanos accounted for 75 percent.[45] These remarks, however, are a reversal of the definition of a minority group, which is partly characterized by a difficult access to the same opportunities than those of the dominant group, including managerial rather than subordinate positions such as machinist and technician. As Charles Montaño testifies, "As a minority, no matter what I did, it was never enough."[46]

Native Americans and Hispanos are power minorities, despite accounting for about 60 percent of the state population. In his 1979 thesis on the structural causes of poverty in New Mexico, Nikolaos A. Stergioulas designates the percentage of ethnic groups in the population as one of the main factors, along with education, unemployment, geography, the small size of the local market, and an imbalanced labor market (i.e., an excess of unskilled and semi-skilled labor).[47] One could postulate that immigration from Central and Latin America was a major cause of poverty in the state, but according to the 1990 Census, three-quarters of the poor were racial or ethnic minorities and 90 percent were native US citizens. Native Americans constituted the poorest racial group, with a 1989 poverty rate of 47 percent (27.7 percent for Hispanos and 10.4 percent for the White non-Hispano population).[48] The fact that Stergioulas and other scholars have made the connection between poverty and racial minorities necessarily raises the question of equal opportunity.

Discrimination is also relevant to this study because it is a fundamental element in the establishment of a neocolonial labor system. According to Chicano writer Mario Barrerra, a colonial labor system is defined by the segmentation of the labor force "along ethnic and/or racial lines" and by the systematic maintenance of one segment "in a subordinate position," "disadvantaged with regard to the labor market or labor process in comparison with another group of workers"—in this case domestic migrants from other states. Barrerra identifies five aspects in a colonial labor system: 1. labor repression; 2. the dual wage system between minorities and nonminorities; 3. occupational stratification or the "practice of classifying certain kinds of jobs as suited for minorities and others as suited for non-minorities"; 4. minorities as a reserve labor pool to give elasticity to the labor force; and 5. the "buffer" or "shock absorber" role that allows laying off minority workers disproportionately.[49]

In the twentieth century, most of these aspects had been eliminated, but others continued to be observable, which justifies the use of the term *neocolonial.* Occupational stratification at the Laboratories, in the uranium mines, and in the other workplaces of the nuclear chain was maintained, as proven by statistics and accounts of workers. The use of minorities as a reserve labor force was one of the bases of work organization during the Manhattan Project and was perpetuated. Even if no specific records can be found of a discriminatory system within the MED in the wartime period, historian Peter Hales identifies a policy based on six directives that match Barrerra's criteria: 1. hire colored workers reluctantly; 2. use such workers only for low-skilled, low-paid jobs; 3. "observe the local status quo as long as that status quo militates against integration and improved conditions for colored workers"; 4. espouse separate but equal; 5. limit contact between white and colored; and 6. prevent the creation of a community.

These rules primarily applied to African American workers at the Hanford and Oak Ridge sites, but in New Mexico "colored workers" were those who the District called "Indians" and "Mexicans," everyone "from Navajo and Pueblo traditionalists to Castilian Spanish-Americans."[50] Lab workers described the "buffer" role they thought they were assigned when saying they felt dispensable. Even further, Gregory Cajete of Santa Clara thought they had been chosen to be the neighbors of a nuclear weapons laboratory because "if something did indeed go wrong it wouldn't affect too many people," and the people it would affect were considered "almost expendable."[51] The lawsuits filed against the Lab for discrimination during the lay-off waves of the 1980s and 1990s are proof of that role as well. And, as a telling example of espousing separate but equal, "Hispanics and Native Americans often told of not being allowed to eat in the same lunchrooms with whites and being referred to as 'aborigines.'"[52]

For the first decades, the lack of Hispanos or Native Americans in scientific positions was regarded as "as a function of educational level rather than systematic disorder."[53] Years later, however, as interviews suggest, New Mexican graduates of local tertiary institutions were still unable to compete with outsiders. A glass ceiling came to exist, which prevented local non-Anglo residents from obtaining upper staff positions. Loyda Martínez, who started working at the Lab at age eighteen as a secretary through the vocational school, announced her will to get involved in politics in Washington, so she could come back to "help her people." She explains,

> When you reach a peak at the Lab as a minority, you hit this glass ceiling and many people are disillusioned. I have got the ambition, the motivation, the drive, and the education, but I never got the fairness and the opportunity that my white counterparts have. We need more minority role models to be a drive for younger generations. Because of socio-economic barriers, we don't get the opportunities.[54]

Her discourse addresses preferential treatment that reveals various forms of ethnic, geographic, and educational discrimination, while also reflecting the growing militancy of younger generations.

At the same time, many struggled to balance the overall benefit of Los Alamos with its faults. Paul Frésquez heard stories about minorities being kept down from positions such as group or division leader because "they were always not qualified enough or something to that effect, when they had been working there quite a long time." Still, he emphasizes what his community owed to Los Alamos economically, calling it "the heart of the northern community." Without it, people wouldn't be "as well-off as they are" and "we would have a ghost town in Española," he said.[55] This reflects the deeply rooted fear, inspired by the boom-and-bust history of the West, of being deserted by the providers of economic activity. The paralyzing effect of such fears made it a difficult process to recognize and denounce the mechanisms of neocolonialism. The debate on whether to support or condemn the institution was a central discussion at every level of society, sometimes turning into an intergenerational argument. Pueblo member Kaa Fedeh relates that his father considered the Lab a blessing. Conversely, his son, freshly graduated from college, affirmed he would never work there. Some Lab critics believe promises never materialized as Native Americans and the Hispanos were still "at the bottom of the work ladder."[56] One Hispano mentioned how proud he felt of his son coming to work with him until "he got older, and started thinking about what I did, what Los Alamos was all about, he stopped coming with me. . . . That hurt. Los Alamos is an awful place to live. . . . It's expensive. And they treat you like second class citizens."[57]

Others took action to voice their sentiments of injustice. Rubén Montoya, who had worked at the lab since 1946, wrote a letter to Senator Joseph Montoya in 1973, in which he states,

I have been a victim of discrimination. Mr. H., section leader was responsible for most of the harassment. If someone went to that section, they would see that Anglos always got better jobs and raises than Hispanic surname persons, no matter how qualified they were. You wouldn't find any Anglo laborer or janitor or any other menial jobs: this kind of work was left to us, so-called Mexicans.[58]

He also claimed he had been asked to look for another job after hurting his back in 1971, while Anglo colleagues, who were injured at work, were better treated. Montoya's testimony suggests that, although education and skill are repeatedly cited as the path upward on the social ladder, discriminatory practices inherited from the past and produced by a culture of excellence need to be factored in to explain why the mechanism seems jammed. The victimization of the people who were put in inferior positions is a parameter that should not be overlooked, yet, no matter the degree of exaggeration or bias in these workers' testimonies, socioeconomic figures support their experiences in portraying Los Alamos as an island of privilege in a state that has failed to capitalize fairly on its nuclear boon.

Cultural Dilemmas: Reconciling Traditions with Modernity

In 1950, half of New Mexicans lived in cities for the first time and four thousand farms went out of production during the decade. The demographic shift from a rural, agrarian culture to an urban, industrial one was accompanied by the increase in welfare money that agricultural counties received. Nine counties benefited from job and population growth, while in sixteen declining counties, where federal spending was close to nonexistent, welfare and social security payments increased exponentially.[59] As New Mexicans sought to better themselves through more profitable jobs in the nuclear economy and traditional occupations withered, substantial cultural changes occurred. In Hispano communities, the disappearance of former lifestyles caused the loss of ancestral know-how, which was then criticized by traditionalists.

In the 1960s, New Mexico Hispanos, led by the Alianza Federal de Mercedes, headed by Reies López Tijerina or "King Tiger," began asserting their historic rights to the land conferred to them by the Treaty of Guadalupe Hidalgo.[60] Their fight revived ancient struggles and emphasized the dynamics of neocolonial relations between ruling and labor classes. The reactions spurred by the raid against the Rio Arriba County Courthouse led by Tijeras in 1967[61] were "typical of the responses by dominant groups everywhere to the mutinous rumblings of the subjugated."[62] This social movement put Hispano workers from the Valley employed at LASL in an awkward position: "They were regarded in varying degrees as *vendidos*, or sellouts, . . . trapped between the increasing militancy of Hispanos and their own psychological and material aspirations in the Anglo world of Los Alamos."[63] Some found it difficult to defend a cause that no longer concerned them directly or in which they were no longer seen as legitimate partakers. Then, revelations on the release of radioactive waste into the environment further antagonized the people living around nuclear facilities. Accepting a job there no longer solely meant seizing a golden opportunity, it was now also regarded as supporting the institution that had brought alarming health and environmental issues to the region. These sites became symbols of both social success and cultural betrayal. Local workers were torn between two worlds: one they were attached to but could no longer provide sustenance, and the other that they had come to despise despite its advantages.

In the 1990s, concerns for cultural preservation surfaced. Arellano writes, "*El que pierde su tierra pierde su memoria*,[64] and no amount of money or technological advances will help us recover that loss." While some have considered Los Alamos the "Promised Land," others see it as "a virus destroying all the data on the hard disk." In Arellano's view, to preserve Hispano culture, New Mexicans need "a strong rural economy." The viability of a strong rural economy in New Mexico may be considered wishful thinking in the present day, considering the plight of rural farming–dependent counties such as De Baca, Harding, and Union.[65] Culturally, however, the split between traditionalists and modernists is significant because it symbolizes New Mexico's identity crisis. A new model of community economic development emerged in the Hispano villages "to replace the region's impoverished colonial economy with a bioregional one: a diversified, resilient economy that is both culturally appropriate and environmentally sustainable."[66]

Arellano also designates the situation in the Rio Arriba Bioregion as a colonial economy. He blames Los Alamos for introducing an "economy based on a fantasy" maintained through the infusion of material frustrations. While locals strive to "get past a certain wage level," the new protectors of the land are antinuclear environmentalists who do not share the region's agricultural past.[67] A past regretted by some, like Paul Montoya, who was seven when US marshals asked his family to leave their ranch on the Pajarito Plateau. He says, "It was a blessing to get out of the work—no more hauling, no more chasing cattle . . . We thought of it as a blessing until we realized there was no more going back." Montoya and his brother got jobs at LANL. He considered himself lucky for the job but losing the land "always lived in his mind." When he retired after thirty-one years at LANL, however, he was diagnosed with beryllium sensitivity, a condition that weakened his immune system, attributable to his work as a fabrication technician handling nuclear materials. His grandson, Gilbert Montoya recalls how "every once in a while, [his grandfather] used to cry just like a baby, and [Gilbert] would say 'What are you crying for?' 'Oh, my ranch,' he said. That sort of emotion and connection to the land left its mark."[68]

Although language has been mentioned both as an indicator of discrimination (for Spanish) and of social betterment (for English), it is also a key indicator of cultural change. In hindsight Leroy Martínez, a mechanical engineer at LANL, regrets that the tradition of speaking Spanish is fading. Upon seeing the people coming from all over the country, he notes, "It seems like they don't have any roots, they don't have any traditions. They don't know where their parents and grandparents came from. I can pinpoint everything in the network of families that we have here." On the one hand, Leroy felt traditions were fading because of Los Alamos; on the other hand, his job there, somewhat paradoxically, was what enabled him to stay close to his roots and maintain these traditions.[69] Lucille Sanchez says speaking Spanish was discouraged at school and at work. She once worried about having an accent, but now she insists she is proud of her linguistic and cultural heritage.[70] This reversal from rejection to reappropriation of local cultures may also stem from the fact that, as New Mexicans crossed to the other side, some were influenced by the viewpoint of expatriates who saw local cultures through romanticizing lenses.

Some young Hispanos regards their cultural heritage as a way out of the aseptic world on the Hill. One woman of Hispano ancestry who grew up in Los Alamos recalls the first time she went down to the Valley and became aware of her cultural loss after her family's assimilation into the Hill's culture. She says, "I couldn't believe it. I saw cows. A horse. And poor people. . . . In Los Alamos there was no culture, no Hispanic culture.

I had no sense of roots . . . or a home. . . . I didn't even know my mother spoke Spanish until my dad died and she moved away from Los Alamos." The culture shock made her see her hometown in a different light, as "a very materialist place."[71] Local workers had to partly let go of their cultural identity to enter the sterile environment on the Hill and, in turn, this environment modified their relation to culture.

For the students from the Española Valley High School, one striking feature in their caricatured rendition of the Anglo was condescension, an attitude more overtly displayed toward the Native American than the Hispano. The authors depict "some transients, who sincerely believe that without them, Tewa tradition will die" and who "see it as their sacred responsibility to preserve on film as much aborigine culture as possible." Some people believe themselves to be experts who "know more about the Native American than the Tewa himself. Archeologists and West Texas tourists have been observed lecturing Tewas on the meaning of being Native American." The one thing that Tewas and Hispanos have in common, however, is that they "have been unwilling to consent to cultural suicide, their suicide, so that Española can progress towards Americanization" but "in a desperate effort to purge themselves from their feelings of inferiority," they have still tried to emulate their Anglo neighbors.[72]

While different dynamics can be observed concerning Native Americans, occupational opportunities have had a similar dividing effect. The dilemma is often between economic and cultural survival. The arrival of nuclear jobs accelerated the breakdown of traditional Pueblo insularity. These communities suffered from the fragile equilibrium between traditional ways and the intrusion of "modernity" into their organizations. A study by the New Mexico State Planning Office in 1969 reported, "Many Indians feel 'split down the middle,' with one foot in Indian culture and one in the modern world." One Pueblo member sketched a man split in two: one "half with feathers in his hair, arm bands, moccasins, dance kilt, and in his hand evergreen boughs, and flanked by a typical *horno*" and the other half with a "Stetson hat, cigar in the corner of the mouth, shirt and necktie, trousers, and at his side an automobile, refrigerator, money bags, and in his hand a book."[73]

One Native American technician at LANL talked about driving every morning into "the white world" that "is out of control," and returning every evening to "his world," his "land," and his "place of traditions."

He described "a window at his lab overlook[ing] a mesa top which held a defunct reactor next to some old Indian ruins. It was a 'strange' juxtaposition. He felt as though his land had been 'invaded,' and he was concerned about the 'ecology' of the area." Although he was aware of LANL's history and mission, he "avoided thinking about it." Another testifies, "Some people in the Pueblo have found it hard to follow, the changes coming so quick. But it is what it is. Jobs, money. Some don't like it. But you have to cope."[74] Resignation and bitterness as a result of economic pressure often characterized the workers' relationships to their workplace.

Hispanos and Native Americans are sometimes considered as "bicultural" groups, but biculturalism has not necessarily been portrayed as richness. Rather, "cultural resistance" has been considered a factor of poverty.[75] The 1969 study, for instance, divided Pueblo Indians into three categories according to their degree of openness: "conservative, transitional, or advanced."[76] The persistent exodus from the pueblos had many effects on their structures. Those who left the reservations were usually the best-educated youths, and their communities were hence deprived of the leadership they could have provided. Charles H. Lange addressed the phenomenon in the late 1950s at Cochití pueblo:

> Working conditions and employer-employee relationships are out-side the control of pueblo officials, thereby eliminating many forms of economic sanction formerly levied against those who deviated from the main cultural stream. Community-labor projects and tribal offices must be borne by fewer individuals . . . . If families seriously need financial aid, the absentee wage-earners are most likely to help. Pueblo community responsibility is thereby displaced by competitive considerations. Pueblo residents visiting the city homes of these absentees become envious of various conveniences and want the same, or comparable, improvements in their pueblo homes.[77]

Gender roles were also affected. After 1943, men transferred their agricultural skills to whatever assignment they could get on the Hill, and many Pueblo women were hired as maids. They left their children in the care of grandmothers, while they went to look after the scientists' children. The growing demand for housekeepers made women more valuable than men. They sometimes earned a better salary. The direct consequences were that fields were left uncultivated, and men were stripped of their traditional roles as providers. Thus, the pueblo became notably more dependent on the outside world for its survival. Cultural arts and crafts were also increasingly mass-produced to meet the demand of visitors.[78]

Furthermore, the presence of the Lab had a very practical impact on Pueblo culture because they no longer had access to areas where they formerly gathered wood, water, or greens for their dances. Peggy Pond Church mentions a fence that symbolized the oxymoronic cohabitation of the sacred and the dangerous: on one side a sign bore "in enormous red letters the warning DANGER! PELIGROSO!" and on the other, a "sacred area" had been set aside for Pueblo Indians to "tend traditional shrines and place prayer plumes."[79] Access to sacred areas had posed a problem since the onset of the Manhattan Project. Elmo R. Morgan, a former worker with the Army and the Zia Company, recalls having to settle an argument between the security patrol guards and "two or three Indians inside of the fenced area just conducting rituals or meditating, or whatever, on some of their old, old ritual grounds there."[80] Disregard for Native cultures is made transparent in the phrase "meditating, or whatever." They were regarded as inconsequential.

Governor Walter Dasheno of Santa Clara expresses his indignation concerning LANL setting off explosions or dumping nuclear waste on sacred ground, saying, "The affront to our culture and religion is complete. We should not be required to specify in measurable terms why a sacred area is sacred." In addition to regretting the loss of ancestral lifestyles and know-how like Hispanos, Native Americans deplore the loss of spiritual legacies through the destruction of areas used for ceremonial purposes. Geronima Cruz Montoya of Ohkay Owingeh affirms, "The disappearance of such sanctuaries has left a vacuum which nothing the white man has to offer will fill."[81] Outrage created new resistance strategies. Today, several cultures of preservation cohabit in New Mexico where the providers of nuclear jobs are juxtaposed to traditional communities. Each culture attempts to protect itself from the attacks of the other while they continue to permeate each other nonetheless. Though they are more diluted in the larger cities, these cultural tensions are a statewide phenomenon. They are the cultural dilemmas that New Mexicans confront when they seek to enter the dominant culture presented as the most legitimate way of life and worldview.

Eleven

Land, Lawsuits, and Waste

Due to the Cold War, all information more or less related to nuclear weapons was automatically classified. According to historian Garry Wills, society was under the executive control of "a National Security State" and "pervasively militarized" from World War II to the war on terror. He calls this period "two-thirds of a century of war in peace."[1] The most disturbing fact about this era and the apparatus of confidentiality utilized by the state is that its purpose was not solely to protect the nuclear arsenal from outside enemies, but also to conceal its negative effects and protect the complex from the American public. Eileen Welsome writes in *The Plutonium Files*, "Much of the secrecy was prompted by fear of lawsuits and adverse public relations."[2] In the 1930s, the deaths of watch factory workers, who painted the dials with radium and inadvertently ingested it, had triggered lawsuits and worldwide public sympathy. Thereafter, litigation was seen as a potential threat to government nuclear projects. The fear turned out to be justified.

Land Still at the Heart of Discord

Unsurprisingly, land controversies figure among the first instances of militancy in the state. This longstanding, unresolved issue at the core of the region's roots was still the center of attention at the turn of the twenty-first century. First, in the 1970s and 1980s, some of the 150 or so expropriated ranching families in White Sands fought against the federal government and the Army. When the government first took their land and grazing rights in the 1940s, some kept their leases and were subleasing to the government. When the range expanded eastward in 1955, some of those who had relocated had to give up part of their properties again. Then Congress passed several acts to acquire the land definitively.[3]

One rancher aforementioned, John Prather, took a stand against the expansion. The Army was unable to gain control over his land until his death in 1965. Military officers offered him $200,000 for his property and warned him that missiles would be launched over his house, but he continued to send the check back, refusing to cash it and saying, "I'm going to die at home." The White Sands authorities rejected his symbolic offer to lease his ranch for $1 for ninety-nine years. Irving Porter, a friend of the family, called Prather "a thorn in [the Army's] side." About twenty-five ranchers came to help with food supplies and transformed the main house into "headquarters." The old man was portrayed in the media holding a gun, in a state of siege, a prototypical image of a cowboy defying the authorities in the West. In the end, Prather was granted fifteen acres of land around the house after he had gone back to court and obtained a writ exempting the area from confiscation. The rest of his property was incorporated into the military reservation. After his death, the Army sent out a group of representatives to his funeral while seizing the remainder of his house.[4] John Prather became a symbol of resistance in the local and national media and inspired other New Mexico landowners a few years later.

The government announced that ranches would not be returned, despite the signed agreements stipulating that all personal property would be restored, and ranchers would be compensated for their losses. The condemnation proceedings provided that the landowners would be paid only for the land they actually held in deed. The ranchers felt cheated and organized in the White Sands Ranchers Association to seek additional compensation in a Court of Claims lawsuit in 1969. New Mexico Republican Congressman Joe Skeen, who testified on their behalf, told the House Judiciary Subcommittee on Administrative Law and Governmental Relations, "It is ironic that a nation which can establish a Marshall plan to rebuild Europe after World War II has failed to adequately compensate its own citizens." Skeen was accompanied by two ranchers, Alyce Cox and G. B. Oliver. The former underscored the fact many of these families had members in the service during the war and "there was no reason for the ranchers to be treated in the rude and dictatorial manner by these employees of the government." The latter claimed the ranchers had been "held hostage by one kind of lease or another for over 40 years."[5]

The legal battle extended for more than twenty years and resulted in the White Sands Fair Compensation Act of 1989, sponsored by Senator Pete Domenici, for which $17.5 million was set aside for additional compensation, but the government reneged, claiming that it would need the funds for defense costs.[6] A few local residents found another way to make money from neighboring the range. Kathy and Oliver Lee allowed missile firings on their ranch about thirty miles north of the range's northern boundary. The Army paid them a flat fee of $10,000, plus $2,500 for each test. But these practices were criticized as another form of extension of the range's activities onto private lands and they presented security issues: one tactical missile fired from Fort Bliss toward the Lees' ranch went awry because it suffered a computer software error and landed twenty-five miles short of its target on another private ranch.[7]

The 1980s saw the legal battles heat up. In 1982, Dave MacDonald, eighty-one-years old, and his niece Mary became nationally famous for their armed occupation of their family ranch within the boundaries of WSMR, refusing to let go of their property. They put up barbed wire on the road and held up signs reading "road closed to U.S. Army" or "deeded land no trespassing." By that time, the Army had acquired about 70 percent of the private lands either through arrangements or condemnation. McDonald declared, "It looks like they just want to age me out." Many other aging ranchers had simply sold out at whatever price the military set for their land and grazing rights, which amounted, in McDonald's case, to $30,000 (an estate agent had appraised his lease holdings at $1,600,100).[8] That same year, agricultural economists John Fowler and James Gray at New Mexico State University prepared a staff report in which they figured that the cumulated net income plus interest from a one thousand-head ranch operated in WSMR between 1942 and 1982 inflated to 1982 dollars was about $10 million. The forgone net income for McDonald's ranch after subtracting the government's lease payment was $4.9 million.[9]

At the fortieth anniversary of the Trinity blast, Dave MacDonald and seventeen other displaced ranchers and their family members gathered outside the Tularosa gate to WSMR to remind visitors heading to the site that the local population was still awaiting compensation. Mary McDonald declared, "The government lied, cheated, and stole from those people who gave up their land."[10] The fight died out when the ranchers' children no longer had the motivation or finances to pursue it. Dave McDonald paid out considerable legal fees, but the lawsuits were unsuccessful. In the end, the family received $60,000.[11] The McDonald Ranch is now part of the Trinity tour as a historical landmark and was restored for $256,000 before being opened to visitors who can see an exhibit on the last preparations of the "Gadget."[12]

Environmental historian Ryan Edgington has explored the convoluted fifty-year struggle among the military, ranchers, environmentalists, and state and federal environmental agencies to gain control of the White Sands military reserve. He explains that the government "found themselves wrestling with the politics of the old and new West" and calls McDonald "a folk hero of sorts."[13] Like Prather, he was the symbol of a dying myth. But military presence and the loss of private properties did not mean the end of life in the Tularosa Basin according to Edgington, who argues that removing livestock and barbed wire "has transformed a rural landscape once dominated by small ranches and an extensive cattle business into an unexpected haven for wildlife."[14] It is interesting that Ernest Aguayo recalls that in 1994–1995, during a visit to his former ranch, he saw wild horses, wild cattle, and oryx, an African import, roaming the range.[15] The last cycle of conquest was thus led by nature, which reclaimed its rights over the area.

More protest has been rumbling among descendants of families who lived in the Basin at the time of the Trinity blast. In 2005, they founded the Tularosa Basin Downwinders Consortium to draw attention to the high number of radiation-related diseases in the communities affected by fallout from the test. They blame the government for having been "negligent" and never having "undertaken an epidemiological study of what happened subsequent to the test." The association reports cases of miscarriages, effects on eyesight, internal ingestion of contaminated food, milk, and water, and many different kinds of cancers. According to Tina Córdova, cofounder of the consortium, she, her family, and the other members of her village "were unknowing, unwilling, uncompensated participants in the world's largest scientific experiment with devastating consequence."[16] Their militancy proves that the political fallout of Trinity is not yet an issue of the past. On July 31, 2015, New Mexico Senator Tom Udall made a speech at the US Senate on the behalf of Trinity downwinders and asked to include them in the 1990 Radiation Exposure Compensation Act (RECA). Udall mentioned Henry Herrera, eleven years old in 1945 and a cancer survivor, who told him, "I am the only one alive to tell about it. Everyone else has died of cancer." The oldest daughter of Edna Hinkle, another victim, told him, "It is not a matter of *if* you get cancer it's a matter of *when*."[17] That summer, the protesters perpetuated the tradition of demonstrating at the site for the seventieth anniversary of the test to attract media coverage.

The second major land controversy that extended over the same period concerns the evicted Hispanic homesteaders of the Pajarito Plateau. In the 1960s, Tijerina, who had launched the fight for the return of land grants to their original Hispano owners, did not spare LASL in his attacks. In 1969, he attempted to make a citizen's arrest on Norris Bradbury, then laboratory director, for trespassing on the Ramon Vigil Land Grant.[18] One can say the homesteaders' mobilization was an offshoot of the land grants movement. At the end of the 1990s, when they called for fair compensation for the land they had given up to the government, Alianza members seized the opportunity to make renewed claims of their own. But the Guadalupe-Hidalgo Treaty Land Claims Act, while having passed the US House in September 1998, never passed the Senate. President Bill Clinton issued a statement of administration policy saying, "This bill would renew land title disputes that already have been resolved by an international agreement or operation of law, in many cases over 50 years ago."[19]

The fight was taken up by the descendants of those who lived on the Pajarito Plateau before the winter of 1942–1943. Contrary to White Sands, the issue in Los Alamos was complicated by the number of claims as a result of the different property systems that had succeeded each other: the indigenous, the Spanish, the Mexican, and the American systems. Representatives of these groups each claimed to be holding more legitimate rights to the land. San Ildefonso Pueblo had been pursuing a claim on the plateau since the 1960s on the grounds of an oral history that "documents a gift of land in late 1942 from the Pueblo to the Manhattan Project to help in the war effort, land that Pueblo officials believed would be returned after the war."[20] In 1995, Senators Pete Domenici and Jeff Bingaman proposed legislation for a transfer of excess land from LANL to Los Alamos County and San Ildefonso Pueblo. Two years later, Congress approved legislation to turn over 4,600 acres of LANL land. The discussion prompted the reactions of the Rio Arriba, Santa Fe, and Española city councils who "called on the DOE to return the land to its original owners or their heirs." After requesting an investigation by the Army Corps of Engineers, "the Army determined that the homesteaders had no legal basis for more compensation, ruling that the original condemnation had been proper."[21]

The protesters organized the Pajarito Plateau Homesteaders Association led by executive director Joe Gutierrez, a longtime employee of LANL. In 2000, Gutierrez made a parallel between the displaced victims of the Cerro Grande fire disaster and the homesteaders; he said that, contrary to the former, the latter "didn't have the benefit of media coverage or community donations, nor were there any politicians tripping over themselves to help."[22] The lawsuit lasted until 2004 when Congress approved $10 million for a Pajarito Plateau Homesteaders Compensation Fund to be divided among more than three hundred descendants. Some believed the sum was not enough; others had hoped to get the properties back, but most were tired of waiting—only one homesteader, who had experienced the buyout, Emelina Grant, in her nineties, was still alive—so they approved the deal. Juan Antonio Martínez said the news "took a toll" on his grandfather who had lost three hundred acres and had worked as a janitor at Los Alamos thereafter. On the other hand, he was glad "for the closure of it" and his mother "could use the money." Joe E. Gonzales, eighty-six at the time, said he "missed this place even now" and that he lived in Pojoaque "while 'rich people' live in expensive houses on his family's lost acreage." He planned to spend the money on college tuition for his grandchildren.[23]

A year later, the Pueblo de San Ildefonso Claims Settlement Act put an end to conflict with the Pueblo, which received a payment of $7 million and the opportunity to acquire about 7,600 acres from the Santa Fe National Forest in exchange for letting go of their claims to the Pajarito Plateau.[24] Legal battles between the government and New Mexicans often ended in the same way: with a settlement that would quiet the protest and provide some measure of economic relief to the families. Legal developments did little to change the relationships between the giants of the nuclear industry and the poorer local populations, but it proved that resistance could attract nationwide attention to issues that would tarnish the image of the industry. *Los Alamos Monitor* editor Ralph Damiani wrote in an editorial, "To take 65 years to pay those whose land was taken is shameful beyond words." This was a time "when our government does not do us proud."[25] However, by casting a new light on the told and retold story of

Los Alamos, the land controversy provided the opportunity to tell the story yet again, this time from a local angle. The homesteaders' fight compelled researchers from the University of New Mexico's Oral History Program to seek "individuals who would be willing to share their remembrances and perspectives in interviews."[26]

The discriminatory nature of the land-grab was an underlying element in the homesteaders' call for compensations. Testimonies surfaced then, disclosing some disturbing stories of Hispanos removed from their lands who were allegedly "subjected to slave-like labor conditions, detention under armed guards, and involuntary medical experimentation." These revelations led to the filing of a second lawsuit by Sylvia Molina, an heir of José Gómez, who claimed he had been required to clean areas believed to be contaminated by radiation. He was regularly examined by doctors and was forced to drink an unidentified substance every day before leaving work. According to Joe Gutierrez, the main purpose of this parallel action was to "strengthen plaintiffs' claims in the land expropriation case."[27] Yet, employees soon began to file new lawsuits for racial discrimination against LANL, which had to make additional settlements for the tension to abate.[28]

The Labs under Pressure

After the end of the Cold War, disarmament meant that there would be less money for the Labs. The Clinton administration created Life Extension Programs (LEPs) to ensure the safety and reliability of the aging stockpile and Weapons Dismantlement programs.[29] The research centers diversified and redirected their activities toward nonweapon fields; nevertheless, the weapons program still accounted for more than half of the LANL budget in 2012.[30] As each weapon or space program was discontinued, either another branch would absorb transfers or employees would be let go. Discrimination cases were filed because Hispanos were disproportionately targeted by the cuts. Federal contracts and subcontractors are required by law "to guarantee equal employment opportunity without regard to race, color, gender, religion, national origin, disability or veteran status and to meet affirmative action obligations," but in 1995, the Department of Labor's Office of Federal Contract Compliance Programs found that LANL had "failed to follow its own layoff criteria, which resulted in Hispanic employees being terminated at a statistically significant higher rate." They reached a $625,000 settlement with LANL in 1998. The money was used in pay-back to the ninety-eight Hispano employees in question and LANL rein-stated twenty-four of them to comparable positions.[31]

Lawsuits multiplied in the early 2000s. In 2003, Randy Padilla of Velarde filed seventeen charges against the Lab with the Equal Employ-ment Commission. He had applied for twenty-four jobs between Novem-ber and July and had gotten only one interview. He believed he had been discriminated against because he was as or more qualified than the hired Anglos. "It's beyond me why a local person can't get a job in their own vicinity when they're well qualified," he said.[32] Two other cases that same year—*Barber et al. v. Regents of the University of California* and *Garcia et al. v. Regents of the University of California*—combined racial with gender discrimination. They alleged "that the Regents of the University of Cal-ifornia discriminated against female and Hispanic employees in terms of pay, promotion, and educational opportunities."[33] One of the women filing suit, Loyda Martínez, declared legal action "appears to be the only way to get them to implement policies and procedures" that are nondiscrimina-tory.[34] The parties in both legal actions reached a settlement agreement including a $12 million settlement fund in 2006.[35]

In 2005, financial auditor and head of the Hispano Round Table Chuck Montaño also sued LANL after spending nine months isolated in a basement, "stripped of all duties after blowing the whistle on what he terms waste, fraud, and abuse at the nuclear facility." The parties settled in 2011. Montaño said he felt he had had a "bull's eye on his back" for the final sixteen of the thirty-two years he had worked at the Lab because he had organized the first employee group. He declared that, although he had an MBA from UNM and had acquired four professional certifica-tions, he was not able to secure advancement at LANL.[36] He was always being "passed over by people with fewer credentials, fewer years of ser-vice, Anglos." Montaño sums up the antagonisms that resulted from the Lab's recruiting rationale that had remained unchanged since its creation. Even though he identified the Reagan era as the period when he "saw less reluctance on the part of lab managers to be blatant in the way they treated minorities: it was no longer taboo to keep shoving them aside. It became chic to be ultra-conservative." The lab had not shed the practice of hiring domestic migrants, making New Mexicans feel as unwanted for-eigners in their own land:

They are being brought in from back East by their relatives or friends already working at the Lab, giving me the message that my degrees from UNM are not as good as theirs. Every day, you are reminded that because you're a native of the area and you went to New Mexico schools, you're not kind of the same caliber as somebody that went to an out-of-state school. It doesn't mean it has to be an Ivy League: from anywhere, you're better off. . . . This is an enclave community of outsiders with one biding overwhelming characteristic: they are not from New Mexico. If you are from within New Mexico, you will always be a visitor here. The people that isolate themselves up there, when they go anywhere else in the state, it is almost like an excursion to a third world country. They identify with each other, with the fact that they are not from New Mexico; up here, they belong. Those of us who are New Mexican and live in Los Alamos, we are the outsiders.[37]

Ten years after the lawsuit, Montaño exposed the Lab's dysfunctions and its neocolonial practices in *Los Alamos, Secret Colony, Hidden Truths: A Whistleblower's Diary.*

These examples underscore the role played by financial compensation in legal matters regarding the nuclear industry. Settlements or the dropping of charges for lack of money to pay legal fees have been classic conclusions to cases involving nuclear labs and corporations. Attorney and author Mike Scarborough, who represented a number of employees between 1980 and 1995 with "lawsuits seeking recourse for discriminatory and retaliatory actions taken against them during their employment," including gender discrimination and exposure to radioactive substances, identified the DOE as the "major culprit" who provided LANL with money to protect its "right to discriminate." He observed that in every case that came to a conclusion, "the firm contracted to represent LANL received considerably more than [his client and he] were awarded."[38]

In addition to these attacks by individuals and organized groups, LANL is under increasing pressure from state regulators to meet deadlines in terms of cleanup. In 2014, they faced penalties because of delays in cleaning up hazardous waste sites that were threatening groundwater. The Lab claimed they did not have the funding to clean it all up at once, which helped them obtain extensions from the Environmental Department; but the

Department stopped granting them.[39] Militant groups now are more active in providing the public with alternative sources of information on the Labs' health and environmental impacts to expose the institution's misdeeds. As evidenced by the rise in attacks against Los Alamos, local populations have fully embraced new possibilities to assert their rights and seek reparation. These manifestations of longstanding underlying tensions are also locals' chance to voice their position on the future of the industry in the state.

In Sandia, employment was the source of polemics as well, only to an extent. SNL had its own whistleblower case in 2005 when Shawn Carpenter filed suit for wrongful termination after having been dismissed for revealing a security breach to the FBI. However, the corporation did not suffer as many accusations regarding racial discrimination as LANL, perhaps because of its positive reputation as a major job provider. Albuquerque mayor Martin Chavez made it clear in 2002 to the University of Texas that he did not want them taking over SNL and even declared, "We are not anybody's colony. . . . We don't want Texas running New Mexico."[40] So while Los Alamos was regarded by some New Mexicans such as Chuck Montaño as a colony of sorts, SNL was put forward as a symbol of independence. A year later, the Hispano Chamber of Commerce gave Sandia its inaugural Aguila Award. Chamber President Loretta Armenta declared for the occasion, "Sandia National Laboratories has been one of the most stable and reliable economic forces in New Mexico for the past 50 years."[41]

Despite a wave of layoffs in the late 2000s, in 2010 the corporation remained a "bedrock of more than 10,000 high-paying local jobs, an 'essential pipeline' of new jobs for local talent." It was still pumping "big bucks into New Mexico's economy to the tune of more than $2 billion a year" and paying $67.5 million in corporate taxes to the state. That year, Sandia hired a little more than seven hundred people and President Paul Hommert proclaimed that, while "many of them are new to this great state," many others "are from our community and educated in our schools." A 2010 impact report, however, showed a minority of 203 among the newly hired had, in fact, graduated from a New Mexico university.[42]

Cleanup was the greatest controversy that Sandia faced in the early 2000s. The Mesa del Sol project, a green, sustainable, and water-wise urban community on the southeast corner of Albuquerque, drew heat from activists because it was to be situated approximately two miles from a mixed-waste landfill, which had been operated by SNL as a "classified landfill" from 1959 to 1988 to dispose of radioactive hazardous waste. Although SNL officials and the NMED believed the repository to be safe, scientists commissioned by Citizen Action, an Albuquerque watchdog group, concluded that Sandia's risk assessment was flawed in various ways such as calculating health risks to adult males without including women, children, and infants.[43] In 2005, the secretary of NMED, Ron Curry, issued the final order allowing SNL to cover the dump with three feet of dirt under the Long-Term Environmental Stewardship Program.[44]

Citizen Action filed a lawsuit against NMED on the grounds that shallow burial of high-level waste is prohibited under federal law.[45] Their final report in 2015 maintained the site was still a threat as it is located above Albuquerque's sole-source aquifer. Environmental lawyer and executive director of Citizen Action Dave McCoy claimed to have evidence that the trickledown effect into the city's water supply had begun. Nickel, chromium, cadmium, and nitrates had started entering the groundwater, and their presence has been documented by the very agency, NMED, that approved the cover solution. Meanwhile, commercial development of Mesa del Sol began in 2005. Senator Jerry Ortíz y Piño asked Sandia to comply with NMED's final order, "which had approved the dirt cover on the condition that Sandia reevaluate the situation every five years to assess whether excavation had become necessary." Eight years later, Sandia had not performed the assessment but had submitted a private request to NMED that this requirement be waived, and it was granted.[46] Reclamation of such polluted sites remains a conundrum and an alarming prospect for populations caught in the middle of jousting matches among government agencies, corporations, and activists.

Radioactive Waste Above and Under Ground

Like the mixed waste landfill, abandoned uranium mines and mills in the northwestern part of the state pose a serious environmental threat even as the area is coveted for renewed exploitation. In his speech to defend the plight of the Tularosa downwinders in 2015, Senator Tom Udall also asked for the post-1971 uranium workers to be included in the RECA legislation and to "fund a critical public health study of those who live and work in the uranium development communities."[47] Three out of the fifteen national priorities list or superfund sites in New Mexico are former uranium mining or milling sites: Homestake Mining Company in Milan, United Nuclear Corporation in Church Rock, and the Jackpile-Paguate Uranium Mine at Laguna Pueblo.

At the Homestake site, where the company operated a mill between 1958 and 1981, the DOE had employed about forty Navajo construction workers to consolidate the Shiprock mill tailings in 1985 and 1986, covering the two piles with about seven feet of impermeable soil and three feet of rock. But the cleanup was ineffective. The groundwater around the tailings remained contaminated. Removal activities started in August 2012. The process is defined by the EPA as "short-term cleanup intended to stabilize or clean up a site that poses an imminent and substantial threat to human health or the environment."[48] In Church Rock, where the United Nuclear Corporation operated a mill between 1977 and 1979 when the spill occurred, up to five hundred predominantly Navajo people lived within two miles of the site and used the land for sheep, cattle, and horse grazing. They now use bottled water for drinking because the well water tastes bad.[49]

The Anaconda Copper Company operated the country's largest open-pit uranium mine, which covered three thousand acres, from 1953 until 1982. The Jackpile-Paguate mine is located on Laguna Pueblo land and remained untouched for seven years after the end of operations. Its tanks filled with rainwater, which invited children to swim and animals to drink, and leached toxic byproducts and metals into underground water, until the pueblo started reclamations itself.[50] Studies on fish documented that surface waters in the Rio Paguate and the Paguate Reservoir contained elevated levels of isotopic uranium. In 1986, the Pueblo of Laguna, the Bureau of Land Management, the Bureau of Indian Affairs, and the Anaconda Copper Company entered into an agreement for the site to be cleaned up. In June 1995, the Jackpile Reclamation Project was officially completed. Nevertheless, in September 2007, when a Record of Decision Compliance Assessment was performed, the report concluded that cleanup was incomplete.[51]

Tribal representatives, indigenous activist groups, and grassroots organizations have been fighting for additional and more efficient cleanup procedures on their lands. Dust from the tailings still spreads through wind and water, contaminating soil, plants, and animals, including the livestock that, in turn, contaminates the people who consume the milk and meat.[52]

Children play in the abandoned mines, which are also sometimes used by animals and humans for shelter. Signs warning of danger are few. Buildings are most likely contaminated because the Navajo use earth to build traditional hogans. A further difficulty is explaining radioactivity to residents, as it is both invisible and inodorous. No solution exists to permanently reclaim the mines without leaving the waste there. But the Navajo Nation is asking for removal of these wastes to a safe repository.

In the first decade of the twenty-first century, a "nuclear renaissance" occurred, promoting the use of nuclear energy as the most viable long-term way to palliate energy shortages. The old economic incentives were revived. In 2006 the Global Nuclear Energy Partnership (GNEP) was created to revert the mining and processing of uranium back to the United States after the country had turned to importation of cheaper uranium in the late 1980s. In an effort to gain in autonomy, the United States resumed interest in its own uranium deposits in the southwest and companies began lobbying for the reopening of mines and mills in the four-corner region. Given that most of the mines are on Navajo land, company representatives were sent to convince members of the community to lease their lands.

Five companies had projects to reopen mines and mills in New Mexico in 2015.[53] Some of their projects involved in situ leaching (ISL), also known as solution mining, or in situ recovery (ISR), which means "leaving the ore where it is in the ground, and recovering the minerals from it by dissolving them and pumping the pregnant solution to the surface where the minerals can be recovered." This method does not generate tailings or waste rock, however, "the orebody needs to be permeable to the liquids used, and located so that they do not contaminate groundwater away from the orebody."[54] In 1994, Eastern Navajo Diné Against Uranium Mining (ENDAUM) was founded in response to Hydro Resources's proposed ISL uranium projects. The association believed it would contaminate the waters of Crownpoint and Churchrock. Leona Morgan, coordinator of ENDAUM, describes the tactics of uranium companies to lease land from individuals who had received allotments through the Dawes Act: the uranium companies use a "divide and conquer tactic" to target poor individuals rather than communities and promise them monetary rewards.

There are currently 520 clusters of abandoned mines that include more than one thousand sites where conventional, tailing-producing mining occurred on Navajo land in northwestern New Mexico. Instead of referring to a nuclear fuel *cycle*, ENDAUM refers to a nuclear fuel *chain* because there is no solution to treat or recycle the waste. Crownpoint resident and ENDAUM organizer Mitchell Capitan worked as a groundwater technician for the MobilOil pilot-scale ISL uranium mine and saw that using ISR had not enabled restoration of the polluted water to its original state.[55] Thus, uranium mining represents a junction between ambitions of the future and impacts of the past. Companies touting new, "clean" mining methods are using economic incentives to expand their projects in the region, while the physical remnants of past exploitation are still taking their toll on local residents.

The most critical issue of the nuclear age has been the treatment of nuclear waste generated by both military and civilian uses of nuclear energy. Nuclear powers have produced heaps of toxic, indestructible, and nonrecyclable waste without, for a long time, anticipating ways of safely disposing of it, as national security concerns superseded waste management. The solution preferred by several countries for permanent storage (for at least ten thousand years) is geologic repository. The Waste Isolation Pilot Plant (WIPP) project near Carlsbad gave life to this theory. The radionuclides contained in the byproducts of the industry have half-lives counted in tens of thousands of years. Predicting their behavior on a time scale that defies imagination puts the problem in the realm of science fiction, as semioticians throughout the world are now studying ways to communicate with future generations to warn them about the life-threatening refuse we have produced.

Dubbed WIPP in 1976, the nation's repository for defense-related waste from thirty facilities[56] was meant to dispose of transuranic (TRU)[57] waste in ancient underground salt beds at a depth of 2,150 feet, where the barrels would eventually be encapsulated by the salt. The project has divided New Mexicans for more than four decades. The social and political tensions surrounding WIPP boil down to the very core of the state's nuclear dilemma: the pressure of economic anxieties leading populations to enthusiastically welcome highly risky economic opportunities.

Carlsbad locals played a significant role in the selection of the site. Originally, Kansas was under consideration, but Project Salt Vault near Lyons was undermined by public opposition among other things. In November 1971, the *Albuquerque Journal* featured an article about Kansas's opposition, which was read by New Mexico's senator from Carlsbad, Joe Gant. This spurred interest in hosting the repository in Carlsbad to generate economic activity. In the 1970s, the need for a new influx of capital and investment was dire, since one of the pillars of the region's industry, potash extraction, was struggling under the pressure of Canadian imports. Gant and Walter Gerrells, city mayor at the time, worked together to promote the area. Economic incentives also produced unwavering, committed local boosters of the project. DOE touted the local benefits of WIPP, emphasizing how the influx of new employees would have an effect on "home purchases, groceries and services," but also on "the tax base supporting schools, roads, and other public services" and on job creation "in such areas as office supplies, printing services, and provisions for visitors who come to tour the project."[58] Of the 643 employees, 64 percent were hired locally, an employment level that was maintained after completion (about 800 employees in 2014).[59] According to specialist Chuck McCutcheon, support from Carlsbad's local power structure offered "an interesting twist on the NIMBY syndrome."[60] NIMBY (not in my backyard) and BANANA (build absolutely nothing anywhere near anything) had been used in previous debates concerning the building of potentially dangerous sites near inhabited areas.

Nonetheless, the project's biggest challenge was public resistance. After decades of secret, unilateral decision making by government officials and alarming revelations on the dangers of radioactive materials, convincing the public of the viability of an underground landfill for radioactive nuclear waste was bound to be an ordeal. Over several decades, scientists at SNL, who had been given the technical responsibility for developing the repository, toiled on the technical challenges presented by the project, such as the presence of an underground river that flows above the storage rooms and empties in the Pecos River, or the fear that a driller searching for oil or gas might puncture the site somehow in the future. Throughout this period, facts and findings were manipulated by all parties to serve their philosophies, whether it was to keep the project rolling or to stop it altogether.[61]

On the opponents' side, the Pueblo Indians declared their hostility toward the site from the onset because the route the waste would take from LANL to WIPP might go through their lands. In 1982, the Eight Northern Indian Pueblos Council (members of the Nambé, Picuris, Pojoaque, Santa Clara, Ohkay Owingeh, Taos, Tesuque, and San Ildefonso Pueblo Indians) adopted a resolution stating its unanimous opposition to the repository. Gerald Nailor, governor of Picuris Pueblo declared, "They have other areas where they could do this. . . . many New Mexico highways go through tribal lands and the nuclear industry hasn't demonstrated a reliable or safe method of transportation of the waste."[62] In 1991, the New Mexico Alliance, a network of grassroots, environmental, economic, and social justice groups, organized gatherings named Hands United to Stop WIPP in reaction to the designation of Highway 84 as a potential route. The road winds through "rural indigenous, land-based communities such as Anton Chico, El Pueblo, Las Vegas, and Villanueva." In their opinion, WIPP was a way for the production of nuclear weapons to continue and New Mexico should commit to "the pursuit of wiser alternatives for economic development."[63]

A National Academy of Science report in July 1989 affirmed that "the system proposed for transportation of TRU waste to WIPP is safer than that employed for any other hazardous material in the U.S. today and will reduce risk to very low levels."[64] Anticipating the problem, the DOE had begun giving millions of federal dollars to New Mexico for road improvement. Bypass roads were built to divert the trucks away from heavily populated areas. Nevertheless, Don Hancock, WIPP opponent with the Southwest Research and Information Center in Albuquerque, argued that it would be safer for the DOE to leave the waste where it was for the foreseeable future and concentrate on more pressing problems, such as cleanup.[65] Moreover, the record of the DOE in terms of safety on the road was not the most convincing at the time. In 1988, a Radioactive Waste Campaign report found that New Mexico had had thirty-four accidents involving radioactive waste in the previous twelve years. There were no leakages in those incidents, but it remained the highest number of accidents involving DOE waste.[66] New Mexicans worried about the risk of terrorist attacks and accidental spills. Some added that "motorists caught in traffic next to WIPP trucks could be exposed to very slight levels of radiation."[67]

Both in Albuquerque and Santa Fe, residents felt they were to share the burden and risks of WIPP without receiving any of the benefits. Concerned Citizens for Nuclear Safety based their argument on tourism, Santa Fe's economic base, which would be put at risk by the convoys. Owners of restaurants, galleries, and stores, whose businesses depended on visitors, joined in. When the DOE wanted to begin shipment despite the incompletion of the bypass roads, the city council withdrew its agreement to let them pass through, arguing that it did not make any sense to build roads for safety reasons and then act as though they were not needed.[68] Then, in 1999, the first three shipments inadvertently took a shortcut along NM 4, across pueblo sacred land. DOE officials had to apologize to the San Ildefonso Nation for taking the wrong route. Don Hancock used the story to mock the escort's incompetence, saying, "They say they have the best transportation system in the world, and then they can't even find the right route? . . . Give me a break. This is another reason they shouldn't be shipping at all."[69]

In parallel, the debate took place in the political arena. It was a federal-state battle. Governor Bruce King argued that the federal government owed New Mexico the final say over WIPP in view of its historic contributions to nuclear weapons development and research. The DOE eventually agreed "to consider and address the state's concerns before deciding to proceed with construction or bringing waste."[70] In 1982 the Nuclear Waste Policy Act established the necessity of finding a national permanent underground repository for high-level waste by the mid-1990s and gave the states a right to veto the establishment of such a facility within their borders.[71] This kept the debate over possible shipments of high-level wastes to WIPP on the table, even though it had been settled from the start that only low-level wastes would be stored there. The other option was the Yucca Mountain in Nevada, where animosity was just as great against receiving waste it did not produce. While Governor Toney Anaya (1983–1987) fought this possibility, his successor Garrey Carruthers (1987–1991), a Republican close to the New Mexico business community, saw the economics of high-level storage as a way of diversifying the state's ailing economy as in keeping with its nuclear legacy.

Two years after the first shipment, critics such as Don Hancock had not given up, especially once the DOE announced that the site would ultimately also store "hard-core nuclear waste, shipped from power plants around the country and from bomb shops and labs."[72] Negotiations were under way to introduce greater-than-class C (GTCC) low-level radioactive waste at WIPP. To communicate locally on this project, meetings were held throughout the state. Jennifer Richter attended one of these meetings in Carlsbad in 2011. She noted most of the "ardent and enthusiastic support for GTCC waste" came from "local politicians and business leaders, who seemed to speak representatively for the whole community." She argues that, through these meetings, the department transferred responsibility to the public, "making them complicit in the process and also the end result."[73]

But how could the public, who is subjected to intense economic and political pressures, be expected to make decisions whose repercussions on the environment and future generations go far beyond their immediate individual concerns? Not only is the public asked to form an opinion based on complex and often contradictory information, but these meetings appear to only represent the interests of a specific category of the population. So, what is the value in their participation? This new strategy seemed a way to prevent the effects of secrecy by informing people beforehand and thus protect the industry from a posteriori accusations by manipulating knowledge instead of secrets. Ultimately, although the chronology is reversed, the objective and impact are hardly any different.

The heated debate on pursuing profitable, job-creating activities by accepting risks was revived when WIPP suspended operations in February 2014, following a fire involving an underground vehicle and a radiological event, which contaminated "a portion of the mine primarily along the ventilation path from the location of the incident, releasing a small amount of contamination into the environment."[74] Meant to store waste for at least ten thousand years, it was only a decade before this accident occurred at the facility that had taken more than twenty years of research and negotiations to open. The accident, dubbed "a horrific comedy of errors" by a scientific adviser, provided opponents of the nuclear industry with new arguments against its development in the state.[75]

Clearly, the most recent history of the Land of Enchantment remains entangled in its last cycle of conquest. To build more facilities with the consultation of the public, the industry can rely on the deeply embedded nuclearist philosophies of the region. On February 2, 2012, the NRC held another meeting in Hobbs to present the conclusion of its draft report on a proposed depleted uranium deconversion plant.[76] Twenty miles to the south, east of Eunice, a National Enrichment Facility for the enrichment of uranium began operating in 2010. The addition of these new sites shows that the scientific conquest is an ongoing process, making denser the geography of nuclear communities in New Mexico and binding its past with its future. The strains of poverty still play a prominent role in the shaping of people's opinions on the nuclear industry. With a ratio of incomes of the richest to the poorest fifths of households at 9.9, the state ranked number one in income inequality 2008–2010.[77] Those in desperate need for a job will embrace the chance to see new facilities in the area, while others will be able to afford prioritizing considerations about health and the environment and taking action against further development.

Memory

The research upon which this book is based is the fruit of several trips across the Atlantic and many exchanges with New Mexicans, who invariably asked the same question: why would a French doctoral student choose to study the history of New Mexico's nuclear weapons complex and its local consequences? For a long time, my belief was that being an outsider would make my outlook less legitimate, but I have come to believe that being a foreigner helped my perspective be slightly less subject to some biases. On the other hand, there were times when being a local might have helped, especially when confronted with the reluctance of people who did not wish to discuss nuclear issues. In the end, all conversations were valuable, including those that are not quoted here.

During my first stay in New Mexico, one observation struck me profoundly: the marked contrast between the poverty of some segments of the population and the substantial installations inherited from the wartime atomic project. Certain flaws in the official history and literature of the Manhattan Project also surprised me; while the technical, strategic, political, and historical aspects of the bomb as well as the experience of scientists and their families were well documented, references to local populations were scarce. Yet, had these people not taken part in the project too? Some of them had seen their world turned upside down after the arrival of atomic science on the Pajarito Plateau in 1942. They had gone from an agrarian, isolated lifestyle on their homesteads to working on the most futuristic, life-changing experiment mankind had ever known. What happened to the anonymous participants in the birth of the atomic age?

Local memory was the driving force behind this project. My goal was to focus, from a local angle, on the changes that were introduced into New Mexican society, economy, cultures, and environment by events of a global magnitude to give a voice and a place to the residents who were affected in any way by the development of the nuclear weapons complex. Several groups had to be included: displaced populations, local employees at the laboratories, witnesses of the Trinity test, uranium miners and millers, and also antinuclear activists, supporters of the industry, and all who were involved in the debate on nuclear risks and economic growth. In parallel, including a more global assessment of the postwar economic boom (and boon) was necessary to contextualize the events affecting local populations. This panoramic view was essential to weigh both the immense benefits of the weapons industry and its various socioeconomic, environmental, and cultural costs.

Borrowing from the before-and-after approach often used by those who experienced the postwar upheaval revealed that, despite undeniable improvement, New Mexico remained poor and had become one of the states with the highest disparity between the rich and the poor. How could this phenomenon be accounted for? From the start, a first hypothesis was that benefits had been unequally distributed. Whether the situation was due to ill-advised political decisions or unfair employment policies had to be determined. In the end, both seem to be true. The responsibility for New Mexico's current struggles is shared by the employers in the nuclear industry and local political leaders who have been unable to rectify the imbalance. My concern, however, was to identify the origin and explain the mechanism of this imbalance rather than find culprits. Inequalities followed a repetitive pattern that showed a division between local populations, who had been living in New Mexico since before World War II, and domestic migrants from other states who came to seize the opportunities of a thriving industry. When comparing New Mexico's nuclear history with similar patterns of social relations opposing locals to outsiders, colonialism became evident as a theoretical framework. Making the connection between the local legacies of the Manhattan Project and the state's past as a subregion of the colonized American West enhanced the pertinence of such concepts as radioactive colonialism, nuclearism, and scientific conquest.

The region's past shaped both the resilience of the population against conquering efforts and their disposition to accept federal support and risky industries in their quest for economic stability. The state's chronic isolation was a burden, as the internal colony suffered from being cut off from the nation's economic and political centers. Conversely, it was at times a cultural asset that had allowed the perpetuation of languages and traditions before attracting the military project. During World War II and the Cold War, the definition of isolation shifted from forlornness to strategic and valuable remoteness. New Mexico was so far off that it seemed a place where secrets could be buried, and hazardous enterprises could be undertaken in safety from prying enemy eyes as well as from public scrutiny. My objective was not to present an imperialist or Manichean vision of the situation, but to underscore the repetitive occurrence of problems that are characteristic of a colonial situation and of an area struggling to find its way out of the cycle of dominance.

The region's status as an internal colony, founded on its role as a supplier of raw materials to the industrialized East, did not cease, but was renewed through the construction of the "atomic West." During the war, the neocolonial relation was magnified by intensive militarization, a colonial tradition that eased the spread of the military-industrial complex. New Mexico is still faced with the challenges of dependence and New Mexicans are still seeking to be acknowledged for their participation and hardships as a result of experiments conducted on their lands. The state's unique status within the US domestic nuclear empire combines nuclearist and patriotic stances with the revolt of those who felt exploited.

The atomic pioneers' state of mind was also interesting insofar as it was influenced by the iconic narratives, myths, and fantasies through which non-Westerners saw the American West. Moreover, behind their words, one can glimpse the prejudice and misconceptions locals confronted. The Land of the Enchantment's appeal to travelers had been the consequences of isolation, such as the preservation of ancestral cultures perceived as "pristine," the untouched landscape, or the "simpler" way of life that were romantically portrayed in tourist adverts. These had also been the prime criteria in Ashley Pond's selection of the Pajarito Plateau for establishing his boys' school while touting the benefits of living in the great outdoors, far from the noxious urban environment. The imagery that influenced the atomic transplants' view of their surroundings was largely borrowed from tourist brochures, the Western genre, and the national experience of Manifest Destiny.

Such cliché-ridden ideas about the West and its residents laid the foundation for the relationship between atomic pioneers and New Mexicans. On the scientists' side, locals were often viewed as needed manpower and exotic entertainment. Meanwhile the locals' opinion of the invasion concentrated on the material advantages they obtained from working for the Project. Cultural and social interactions sometimes turned into commercial transactions to satisfy the new tourists' cravings for authentic experiences and souvenirs. The invaded therefore perceived the atomic laboratory and the new colonizers in terms of economic opportunity, whereas the invaders perceived locals in terms of cultural diversity that could distract them from the hectic work they were conducting on the Hill, or from the boredom of living in a small, isolated, secret military camp.

One major difficulty throughout this work was to define what it means to be New Mexican or to be an outsider. Who should be considered natives in comparison to postwar migrants? Native Americans and Hispanos of Spanish and Mexican ancestry are easily included as the ancestral cultures in the region, but what of the Anglos living in New Mexico before the war? It seemed relevant that they be included because they had shared the regional experience of the postwar evolution. I do not pretend to have found a resolute answer to the question of what it means to be New Mexican; however, after listening to people express their attachment to the area with words such as "roots," "belonging," or "something in the soil," I became persuaded that the essence of being New Mexican is to be found in the relationships between people and their immediate environment, which, in turn, shapes social and cultural relationships. Defining the special connection New Mexicans have to their environment, therefore, was an important point to make as it transcends the boundaries separating New Mexico's various cultures, and these ties to the land were profoundly modified or dramatically disrupted by nuclear activities.

The addition of new social classes created by the state's development—the scientist elites and the middle-class employees at the nuclear facilities—resulted in a complex map that revived old conflicts and occasioned new disputes such as those between traditionalists and modernists. Nuclear matters have eventually also transcended divisions as all cultures are represented in the rise of political, social, and ecological activism. Be they

Tularosa ranchers, Hispano homesteaders, or Navajo uranium miners, all shared a concern for their survival, their culture, and their land. The move toward militancy was a long process, though, because it occurred in an impoverished region grateful for economic opportunities. Sluggish industrialization, the development of a wage and cash economy, the damages of droughts and depression, and increased federal presence had put the state in an ideal position to undergo a dramatic transformation.

Upon examining local memories and testimonies, a manifest progression is observable in reactions from one generation to the next. The older generation's approach to the industry was enthusiastic and even protective of what was believed to be an economic miracle. Relief and gratitude were expressed at being liberated from the physically demanding work in the fields and from the strain of having to leave one's home to find employment far away. According to the first workers at site Y, the Manhattan Project had bolstered their attachment to their homeland by putting an end to the migratory cycle to find work outside of the state. These migrations were an indicator of how depressed some areas were after the hardships of the 1930s, which had severely affected the agro-pastoral and extraction economy. On the other hand, the opinions of younger generations were affected by their difficulties to succeed within the research centers. The habit of looking to other states for highly educated scientists and engineers out of the best schools in the country made it more difficult for graduates of local institutions to be hired or promoted. Consequently, in spite of the growing level of education among New Mexicans with each new generation, their efforts did not guarantee them access to better jobs. Social reproduction was a key factor in the creation of inequalities within the nuclear economy.

Many testimonies showed ambivalent feelings that fluctuated from a nostalgic view of the past to the will to participate in New Mexico's future as a symbol of modernity and progress. It took up to fifty years for some of the testimonies of displaced inhabitants on the Pajarito Plateau or in the Tularosa Basin to surface. These people and their children expressed some nostalgia for the homesteading and ranching eras, but it is difficult to quantify the extent to which they might have come to idealize their lives prior to the scientific conquest. For what else can one call it when the Army coveted land, acquired it, emptied it, and peopled it with new settlers? Even though the use of eminent domain was warranted by the war, it is now clear that individuals were wronged and suffered in the process. The military installations were not meant to be permanent. Or so believed all those who lost land. And yet, the sites evolved into communities populated with military personnel and scientists. The fact that locals benefited from their presence helped stifle their voices.

In addition, the ideology of nuclearism and the tendency to self-victimize had to be used to analyze discourses. Many praised the nuclear weapons complex while deploring its impacts. Enthused by the prospect of more nuclear jobs at the Labs, at White Sands, at the military bases, in the research centers, in the uranium mines, or later at WIPP, New Mexicans entered something of a devil's bargain in which they saw the immediate economic profit without knowledge or foresight of the risks involved and of the negative socioeconomic and environmental consequences their enthusiasm would entail. The idea to use the devil's or Faustian bargain metaphor as a starting point came directly from the sentiments interviewees voiced of being torn between employment issues and the negative effects of the industry. The concept also evokes the many faces of a problem. From the point of view of Hispanos and Native Americans in the Española Valley, Los Alamos has been a symbol of hope, a goal, an enemy, an ally, and an abnormality. Each individual history was tinged with some degree of moral introspection expressed through accusations, justification, or protestation. Access to information was the parameter that determined the modification of reactions toward contestation, and activism, or, conversely, support of nuclear activities.

The effects of growth concentrated in certain areas, where domestic migrants had settled. Local workers directly benefited from the new situation as well, but at a rate and in proportions that could not allow them to catch up to modern standards for success. My main conclusion after this analysis was that the state did not achieve the prosperity and independence it longed for because its prewar dependence on outside forces did not come to an end in the second half of the twentieth century. Instead, it was transferred to the prime investor and consumer on the nuclear weapons market: the federal government. The whole production chain relied on the government whether it was for funding or for selling its products. The country's foreign policies during the Cold War sent repercussive waves that primarily affected states such as New Mexico because of the extent of its defense-related installations, which had become a mainstay of the local economies. As the prime employers in the region, it became crucial for DOE facilities to maintain their levels of funding. Both individuals and businesses relied on their activities for economic stability. The state's newfound autonomy, therefore, was deceptive. The state's fate depended on how the government waged the Cold War. As long as new weapons were needed for deterrence, the industry was working at full capacity and budgets continued to be revalued upward. The opposite was also possible.

Today, New Mexico and its inhabitants still rely heavily on public funds that finance research centers and programs, including reclamation projects. My argumentation, therefore, puts forward the idea of continuity in New Mexico's history through the continuance of the state's colonial status in the Union without playing down the radical changes that New Mexican society underwent. Although the state no longer holds a bystander role regarding the development of military and nuclear sites within it, since the end of the Cold War, its problems have multiplied. The most effective symbol of the price to pay for the nuclear bonanza is the environmental legacy of the nuclear industry. What is most striking about the map of this legacy is how poorer areas are disproportionally affected. Environmental discrimination was another addition to the state's enduring neocolonial status. Growing awareness of the dangers of nuclear weapons production and nuclear waste generated anxiety verging on paranoia when the number of cancers seemed abnormally high in certain areas adjacent to nuclear installations. Past practices of dumping liquid waste in canyons and burying solid waste in trenches exposed inhabitants to toxic substances. People started associating their health issues with playing in the canyons as children or with their jobs working near contaminants. They turned to their employers for blame.

Current controversies center on the nuclear industry's newest developments and on the perpetual problem of cleaning up the toxic remnants from the last seventy years. Aside from the collective action seeking compensation for the condemned lands, LANL has also been under the pressure of individual and class action lawsuits for discriminatory treatment since the first layoff waves in the 1990s. Hispano and female employees who were fired denounced the fact that minorities had been disproportionally affected by the layoffs. Settlements were reached, bringing temporary respite from conflict, but these scandals tarnished the Labs' image, further antagonizing its opponents and proponents. Local employees also used legal means to force the Labs to do away with discriminatory employment practices inherited from earlier times when local workers were hired and fired seasonally as a flexible workforce. For more than twenty years, the heated debate over WIPP in Carlsbad has caught the attention of all New Mexicans, supporters and detractors of the nuclear industry alike. With the possibility of depositing higher-level radioactive waste and the release of radioactivity in February 2014, the plant more than ever symbolizes the dilemma between risky installations and job guarantees. In addition, WIPP serves as a reminder of the role imparted to the West as a dumping zone for the nation's refuse. So, as the state's colonial and nuclear past blur into a complicated future, some New Mexicans have chosen to support a return to land-based traditions or to become antinuclear activists, while apologists for the nuclear industry call for more sites, more funding, and more nuclear jobs. A spectrum of opinions ranging from radical hostility to unconditional support for the nuclear weapons complex is now identifiable among New Mexicans.

Meanwhile with the seventieth anniversary of Trinity and the establishment of the Manhattan Project National Historical Park in November 2015, the debate revived on how the nuclear age should be remembered and what should be emphasized for visitors at Manhattan Project sites. The discussion centered on determining whether these controversial sites were monuments "to one of the nation's greatest technological achievements" or "to the destructive potential of the human species."[1] One of the main arguments put forward by opponents to the park project—led by the head of the Los Alamos Study Group, Greg Mello—was the fact that the Los Alamos and Oak Ridge sites are still operative within the US nuclear weapons complex. This situation and President Barack Obama's announcement of a $1 trillion modernization program of the nuclear arsenal put political pressure on the park service to adopt a supportive narrative, and the Union of Concerned Scientists even called the park a "Disneyland for weapons of mass destruction."[2] To some extent, the new park was to ride the wave of a new kind of tourism that emerged in New Mexico toward the end of the Cold War: nuclear tourism. This branch consists in visiting landmarks of the nuclear area: bunkers, missile silos, sites of nuclear disasters, and Manhattan Project sites.

According to Nathan Hodge and Sharon Weinberg, authors of *A Nuclear Family Vacation: Travels in the World of Atomic Weaponry*, nuclear tourism is "fueled by a mix of Cold War nostalgia and morbid curiosity." In New Mexico, whose modern geography has been shaped by the weapons complex, Hodge and Weinberg went on a tour of the bomb's birth sites. They describe how, in Los Alamos, "nuclear scientists sometimes hang pictures of their favorite tests above their desks; they can describe, in loving detail, the very personal reasons for their choices." They report that one of them "admitted that he named his son after Ivy Mike, the 1952 hydrogen bomb test." In Albuquerque, where they visited the National Atomic Museum (renamed the National Museum of Nuclear Science and History), they note how the radiation exhibit in the lower hall was meant to "demystify radiation" and that "the museum had to balance two almost contradictory goals: showing that nuclear weapons weren't at all that scary, while also demonstrating that nuclear weapons were terrifying enough to make anyone think twice about using them."[3]

Museums presenting the state's nuclear history are numerous,[4] and they all aim at balancing the positive and negative aspects of nuclear science while promoting the state's involvement in the national nuclear weapons complex. Another standard in several museums is to display replicas of Fat Man and Little Boy; they can be seen in Albuquerque, Los Alamos, White Sands, and at Trinity. In White Sands, a missile park outside the museum allows the visitor to take a stroll amid replicas of all the rockets and missiles that were ever tested at the range, measuring their colossal sizes to one's height. These sites are the places where the memory of the atomic bomb and its legacy were created. Visiting them is a unique experience for those who have been directly affected by the bomb. One example is specifically riveting because it questions the morality of nuclear tourism. Kyôko Hayashi found a way to come to terms with her *Hibakusha* identity by traveling to the bomb's birthplace. In *From Trinity to Trinity*, she writes of her atomic pilgrimage through New Mexico to Los Alamos and to the Trinity test site. She describes her disturbing emotions as she discovers how, in these places, the museums have capitalized on nuclear tourism by selling t-shirts with mushroom clouds on them and pins of Fat Man, the Nagasaki bomb. She calls Trinity the Hibakushas' birthplace because of the kinship she feels with the animals and plants that died on the day of the first atomic blast. She writes, "Standing on the land that speaks no words, I shivered, feeling its pain."[5] With irony and humor, she wonders how the other visitors would react if she took the Geiger counter that was used by one of the officers to show the level of radiation emitted by Trinitite and put it against her body; it would be bound to start screaming as it picked up the radiation emanating from her. The morality of transforming Trinity into a tourist site had already been an issue right after the war when it was suggested the site be made into a national monument. This discussion focused on how significant the site was but also on how appropriate it would be. A *Denver Post* reporter suggested in 1969 that "Trinity was America's guilt symbol, similar to a German concentration camp."[6]

In 1965, Trinity became a national historic landmark and, in 1975, a national historic site. The site opens its gates to visitors twice a year. A lava obelisk sits imposingly in the middle of the desert where the steel tower stood before the blast. In fact, ideas to transform Los Alamos and Trinity into outdoor museums at the end of the war reflected the tendency to associate the West with tourism and view the region as a place where remnants of past pioneering adventures and tokens of America's greatness get fixed in time rather than be turned into symbols of progress. The new beacon of national defense, a sophisticated deadly new technology was at first not seen fit to remain in the desert once the war was over, but the installations were maintained, and tourists now associate the state with their country's military supremacy—in addition to frontier myths and UFOs. Upon visiting the site, one can observe the attitudes of visitors who appear to display either patriotic pride or silent mourning. New Mexican antinuclear protesters and the descendants of families affected by the explosion have also used the site and its opening to the public as a means to draw attention to their cause. All varieties of individuals and groups converge where the atomic bomb was born, be it to deplore its impact or to revere its power.

The construction of memories of the atomic bomb in America has produced many polemics, especially when that memory is staged in exhibits. For instance, the Enola Gay exhibit of 1995 at the Smithsonian National Air and Space Museum in Washington divided veteran and antinuclear crowds to the point where the project was aborted and replaced by a simple display of the B-29 Enola Gay.[7] Similar debates took place in Los Alamos where memory of the town's role in the nuclear era is staged in two museums. The Bradbury Science Museum, named after the Laboratory's second director, is meant to appeal to adults and children alike with interactive colorful exhibits that promote the role of LANL. A few panels in a corner are titled "What does the Lab do to protect its workers and the public from radiological hazards?" and "Restoring contaminated sites."[8] Two antiwar posters focus on the debate over the Japanese bombings next to an impressive billboard, bearing the heading "Why the bomb?," which provides the well-known answer "because it saved many lives." In another corner of the museum, visitors can give their opinions in a public forum. Some write how thankful they are for the bomb and how much fun they had in the museum while others are more introspective: "This is a hard place to visit but I felt I had to. Something dreadful happened here."[9]

Personalities of the Manhattan Project are commemorated on a wall of fame of sorts alongside local workers who played a significant part in the success of the undertaking. These local workers include Beatrice Dasheno Chavarria of Santa Clara Pueblo, who was among the schoolgirls from St. Catherine's Indian School in Santa Fe who were taken to the Hill in 1944 to work as housekeepers in the dorms. She is quoted, saying, "When I found out about the atom bomb after Hiroshima, I really felt sorry for the poor people who died or were hurt during the bombing. I was disappointed to learn that making a bomb was what was being done at Los Alamos." This comment is counterbalanced by Consuelo "Connie" Fulgenzi who came from Las Vegas, New Mexico, with five other women to work at the pass and badge office. She said about her husband, "I was glad the war was over and Larry was coming home safely. He wouldn't be killed in the invasion of Japan." The wall associates Julia Dasheno Roybal, another schoolgirl from St. Catherine's, with the imports of Anglo culture in Los Alamos. The sign reads, "She enjoyed going to the PX to get a Coke and ice cream with the other girls after work and liked going to the movies on weekends."

The Hispano homestead era is also represented by Severo Gonzales, son of Bences Gonzales (mentioned several times in this book), whose grandfather was the original homesteader at Anchor Ranch before it became a testing site. And the Pueblo culture is represented by Angelita Vigil Martínez, niece of potter Maria Martínez, who worked as a housekeeper and babysitter. She is quoted describing the dances on Friday and Saturday nights at Fuller Lodge or at San Ildefonso Pueblo. To illustrate the drastic secrecy measures, the story of Ramon Sanchez, jackhammer operator, bus driver, and delivery truck driver, is presented. Sanchez was subjected to many hours of interrogation after his best friend, Rafael Aguilar, also a truck driver, had lost control of a loaded vehicle and rolled down the hill. When Sanchez could not get Project officials to believe he knew nothing, he was laid off.[10] New Mexican testimonies are thus represented in the exhibit to reinforce the official story and the anecdotes of the Manhattan Project. They help strengthen the paradoxical image of a high-tech laboratory and internationally famous scientists installed in such a place as northern New Mexico.

The Los Alamos History Museum, managed by the Los Alamos Historical Society, retraces the successive settlements on the Pajarito Plateau from the first Keres Pueblo Indians to the arrival of atomic scientists. The society also remembers the streets of the town by creating the Homestead Tour to restore a few signs of the inhabitants of the area prior to 1942. Almost nothing except the pond and a few log buildings, including Fuller Lodge, remains from that period, so the tour is meant to tell the story of the homesteaders and of LARS. One stop on the tour is at the Romero Cabin, which became a symbol of Hispano homesteading on the plateau. Most of Los Alamos now appears as a memorial for the Manhattan Project from the very entrance sign that reads "Los Alamos: where discoveries are made!" to the names of the streets, such as Bathtub Row or Trinity Lane." In 2011, the County of Los Alamos also dedicated two life-size bronze statues of Oppenheimer and Groves on Central Avenue outside Fuller Lodge. Some had wished for such an outcome at Los Alamos after the war, and the eventual continuation of the Lab did not prevent that wish from coming true. Rather, the memory of the Manhattan Project has become support for the Lab's existence. The juxtaposition of historical landmarks and modern buildings is the physical representation of US capacity to superimpose layers of invasion while nostalgically romanticizing past lives and markers of previous presences.

In addition, Los Alamos remains a popular subject for fiction. An example is the 2014 television series *Manhattan*, a drama production for WGN America, focusing on the lives of scientists on the Hill in 1944 and 1945 and shot in Santa Fe. Governor Susana Martinez said at the time, "I'm excited that this series will highlight New Mexico's celebrated history and our amazing, *picturesque* landscapes."[11] Secrets play a major role in the plot as they intensify the tension among characters, especially between husbands and wives. Compartmentalization is recurrently addressed through the rivalry between the thin man group, who worked on the uranium gun-type weapon, and the implosion group, who worked on a mechanism to use plutonium. Steve Terrell of the *Santa Fe New Mexican* writes, "Another underlying theme of *Manhattan* apparently is how foreign and desolate Los Alamos seemed to the people who moved there to work on the bomb."

Actors Rachel Brosnahan and Daniel Stern comment, "You pulled up and it was like a prison camp. . . . This was like landing on the moon."[12] The actors had internalized the impressions scientist pioneers related in their memoirs about how daunting it was to be so far from "civilization" in those "picturesque landscapes."

The relevance of the show is also in the portrayal of featured locals. One episode introduces the character of Mrs. Ortiz, a Native American maid from a nearby pueblo. She and her son Pablo come up to the Hill for the day's work. One scene presents two scientists' wives talking in the kitchen while Mrs. Ortiz vacuums the living room. One of the women complains, "We have to stand at the gate. Meanwhile the Natives come and go as they please. . . . Speaking of Indian, what's your girl doing to the hoover?" And her friend sighs, "I don't know, she probably thinks it's a gaff!" The condescension in the character's answer shows that the expatriates were not portrayed artificially; some paternalist, disrespectful attitudes toward locals are also presented. The comment about access through the gate is a reference to the perception of locals as harmless, invisible servants. Later in the episode, Dr. Winter, a botanist, tries to collect samples to assess the impact of the Trinity test in the vicinity, but she is struggling to draw attention to the importance of the task. The head of the G division, Dr. Isaacs, provides her with the services of "two GIs, as long as they are not assigned to higher priority projects . . . and they are all higher priority, no offense." As Winter tells the two GIs about the necessity to be able to predict the implications of the "device," one of them asks her doubtfully, "Implications on what? On a bunch of coyotes and tumbleweeds?" This dialogue—along with other scenes in the series, including a heated encounter with local ranchers—is an interesting reference to the distorted empty-wasteland trope and to the role of military and political pressures to test the bomb on time with little regard for other matters, such as the impact of fallout.[13] In most other productions on the birth of the atomic age, which include a variety of films, docudramas, an opera, and a theater play,[14] New Mexicans were rarely represented apart to enhance the brilliance and sophistication of the main characters in a story that keeps fascinating and dividing Americans.

As illustrated in these examples, memory is malleable and subjective. The near absence of the local populations in national debates and cultural productions on nuclear history reflects their struggles to escape invisibility.

The construction of memory and significance of events depends on whose memories are considered important additions to history. In New Mexico, both collective and individual memories have been presented in museums and in downtown Los Alamos in an effort to connect the state with national history and promote the scientific successes that were achieved there. Yet, many memories and much of the Manhattan Project's local legacy are still missing in the collective perspective on the nuclear age. Some of these missing voices have brought life to this work and have shed a new, different light on New Mexico's history with atomic science. As new questions and challenges arise in the Land of Nuclear Enchantment,[15] others will be heard.

Notes

Chapter One

1. Harry S. Truman, "Statement by the President Announcing the Use of the A-Bomb at Hiroshima," August 6, 1945, Public Papers Harry S. Truman 1945–1953, Harry S. Truman Presidential Library & Museum, accessed March 22, 2018, https://www.trumanlibrary.org/publicpapers/index.php?pid=100.

2. Jones, *Manhattan, the Army and the Atomic Bomb*, 34.

3. Henry L. Stimson, Secretary of War, "Los Alamos Ranch School Seizure Letter," correspondence to Albert J. Connell, Director of Los Alamos Ranch School, December 1, 1942, accessed April 12, 2014, http://commons.wikimedia.org/wiki/File:Los_Alamos_Ranch_School_Seizure_Letter.jpg.

4. John Manley, "A New Laboratory is Born," in Badash, Hirschfelder, and Broida, eds., *Reminiscences of Los Alamos, 1943–1945*, 32.

5. Hunner, *Inventing Los Alamos*, 38.

6. "'Impact Los Alamos' Symposia: Audience Responses," *New Mexico Historical Review* 72, no. 1 (January 1997): 78.

7. The full title was "Atomic Energy for Military Purposes: The Official Report on the Development of the Atomic Bomb under the Auspices of the United States Government, 1940–1945."

8. See Goldschmidt, *The Atomic Complex*; Gerard H. Clarfield and William M. Wiecek, *Nuclear America*; Jones, *Manhattan, the Army and the Atomic Bomb*; Powaski, *March to Armageddon*; and Herken, *The Winning Weapon*.

9. See Hogan, ed., *Hiroshima in History and Memory*; Maclear, *Beclouded Visions*; Minear, ed., *Hiroshima: Three Witnesses*; Toyofumi Ogura, *Letters from the End of the World*; J. Samuel Walker, *Prompt and Utter Destruction*; Alperovitz,

The Decision to Use the Atomic Bomb; Ham, *Hiroshima Nagasaki*; Hasegawa, *Racing the Enemy*.

10. Badash, Hirschfelder, and Broida, eds., *Reminiscences of Los Alamos, 1943–1945*.

11. Ermenc, ed., *Atomic Bomb Scientists: Memoirs, 1939–1945*; Bethe, *The Road from Los Alamos*; Robert Serber, *Peace and War*.

12. Kelly, ed., *Remembering the Manhattan Project*; Kelly, ed., *The Manhattan Project*. The Los Alamos Historical Society also published *Behind Tall Fences: Stories and Experiences about Los Alamos at Its Beginning*. More recently, two works have also centered on the women who participated in the Manhattan Project: a novel by Tarashea Nesbit, *The Wives of Los Alamos*; and by Denise Kiernan, *The Girls of Atomic City: The Untold Story of the Women Who Helped Win World War II*, which focuses on Oak Ridge.

13. Ferenc M. Szasz and Jon Hunner are two authors who concentrated their research on New Mexico and studied its relationship with science from a local angle. Szasz is celebrated for his historical work on Trinity and contemporary New Mexico, and Hunner was the first historian to publish on the evolution of the Los Alamos military post into an actual town in the postwar years.

14. This perspective is inspired by that of social historians, such as Howard Zinn or Roxanne Dunbar-Ortiz, who have focused on the struggle of often-anonymous actors and groups rather than powerful individuals and government officials.

15. Bartimus and McCartney, *Trinity's Children*, 4.

16. Carlos Vásquez, "Impact Los Alamos: Traditional New Mexico in a High Tech World, Overview of Project and Symposia," *New Mexico Historical Review* 72, no. 1 (January 1997): 4.

17. Scott D. Hughes, "Impact Los Alamos: Managing Editor's Introduction," *New Mexico Historical Review* 72, no. 1 (January 1997): 1.

18. Vásquez, "Impact Los Alamos," 6.

19. Mitchell, *Coyote Nation*, 14, 3.

20. McDonald, Boldt, *The New Mexico Economy*, 10.

21. Alexander, *History of Sandia Corporation*, 3.

22. Karafantis, "Weapons Labs and City Growth," 99, 118.

23. Ross, "Descriptive Petrography of Three Large Granitic Bodies," 5.

24. US Census Bureau, US Department of Commerce, "Poverty: 2008 and 2009," *American Community Survey Briefs*, issued September 2010, accessed March 22, 2018, https://www.census.gov/prod/2010pubs/acsbr09-1.pdf.

25. Dietz, "The Impact of Los Alamos National Laboratory on Northern New Mexico," 50, 41.

26. Lifton and Falk, *Indefensible Weapons*, ix.

27. David Krieger, "Nuclearism and Its Insecurities," in Tehranian, ed., *Worlds Apart*, 109.

28. Site Y had become the Los Alamos Scientific Laboratory in 1947 after it was taken over by the AEC; it was renamed the Los Alamos National Laboratories in 1981.

29. Niklaus and Feldman, *How Safe is New Mexico's Atomic City?*, 11.

30. Price, *The Orphaned Land*, 33.

31. "Nuclear Colonialism," *HOME: Healing Ourselves & Mother Earth* (Reno, NM: HOME), accessed November 10, 2012, http://www.h-o-m-e.org/nuclear-colonialism.html.

32. See Ward and LaDuke, "Native America," 51–78. Vincent J. Intondi explains how African American antinuclear activists made the connection between nuclear technology and colonialism by explaining that because the uranium used in the bombs—before extractive activities developed on Native American lands—originally came from colonial possessions in Africa, such as the Belgian Congo. Also, some activists feared nuclear weapons might be used to support colonial empires. Intondi, *African Americans Against the Bomb: Nuclear Weapons, Colonialism, and the Black Freedom Movement* (Stanford, CA: Stanford University Press, 2015).

33. Nash, *World War II and the West*, 2.

34. See "Federal Road Projects in New Mexico Territory, 1853–1860," in Turrentine, *Wagon Roads West*, 109.

35. White, "It's Your Misfortune and None of My Own," 57–58.

36. Desbuys, *Enchantment and Exploitation*, 163.

37. Physicist and director of the Oak Ridge National Laboratory (ORNL), Alvin Weinberg, was the first to use the expression "a Faustian bargain" to describe nuclear power in an article titled "Social Implications of Nuclear Energy." Alvin Weinberg, interview by Steven H. Stow and Marilyn Z. Mclaughlin, Oak Ridge, TN, March 31, 2003, "Oak Ridge National Laboratories," US Department of Energy, Center for Oak Ridge Oral History, accessed March 22, 2018, http://coroh.oakridgetn.gov/corohfiles/Transcripts_and_photos/ORNL/ORNL_Weinberg_Alvin_transcript.pdf.

38. Rothman, *Devil's Bargains*, 21.

Chapter Two

1. Hales, *Atomic Spaces*, 10.

2. Dunbar-Ortiz, *An Indigenous Peoples' History of the United States*, 1.

3. See Fergusson, *New Mexico, A Pageant of Three Peoples.*

4. DeBuys, *Enchantment and Exploitation*, 8.

5. Genaro Martínez, interview by Peter Malmgren, Chimayó, NM, November 16, 1995, "Impact Los Alamos Project," Oral History Projects and Video Recordings Collection, Albuquerque: Center for Southwest Research, University Libraries, University of New Mexico, Collection MSS821BC, 1984–2006, Box 1, CDs 60–61.

6. "Impact Los Alamos Project," Lebeo Martínez, interview by Dot Waldrip, Albuquerque, November 18, 1995, Box 1, CDs 62–63.

7. "Impact Los Alamos Project," Armanda López Jackson, interview by Peggy Coyne, Española, NM, March 21, 1996, Box 1, CDs 53–54.

8. F. Chris Garcia, "To Get Along or to Go Along? Pluralistic Accommodation versus Progress in New Mexico politics and government," in Etulain, ed., *Contemporary New Mexico, 1940–1990*, 30.

9. Guthrie, *Recognizing Heritage*, 5, 12.

10. Carroll Botts, "Celebrating Zimmerman at 75: Zimmerman Library Artwork," University of New Mexico University Libraries, Albuquerque, accessed March 3, 2014, http://library.unm.edu/zimmerman75/art.php.

11. See Calvin, *Sky Determines.*

12. See Berthier-Foglar, *Les Indiens Pueblo du Nouveau-Mexique.*

13. Spicer, *Cycles of Conquest*, 1, 6.

14. DeBuys, *Enchantment and Exploitation*, 9.

15. Kay, *Chimayó Valley Traditions*, 20. The hymn is quoted in G. Emlen Hall, *Four Leagues of Pecos*, 67. Also see Laura Gómez, *Manifest Destinies*, 7.

16. William DeBuys, "The Sangre de Cristo Mountains," in Weigle, Levine, and Stiver, eds., *Telling New Mexico*, 27.

17. DeBuys and Harris, *River of Traps: A Village Life*, 65.

18. Juan Estevan Arellano, "La Querencia: La Raza Bioregionalism," *New Mexico Historical Review* 72, no. 1 (January 1997): 32. On the Cosmic Race, which was inherited from the mix between Aztec and Spanish cultures, see MacLachlan and Rodríguez O., *The Forging of the Cosmic Race.*

19. DeBuys and Harris, *River of Traps*, 34.

20. Simmons, *New Mexico: An Interpretive History*, 104.

21. Chávez, *An Illustrated History of New Mexico*, 11.

22. Ball, "The U.S. Army in New Mexico, 1848–1886," in Weigle, Levine, and Stiver, eds., *Telling New Mexico*, 179.

23. See Lamar, *The Far Southwest, 1846–1912: A Territorial History*, for a larger perspective on Anglo influence and on the development of the American West after 1890; and Pomeroy, *The American Far West in the Twentieth Century.*

24. Robert J. Torrez, Former State Historian, "New Mexico's Spanish and Mexican Land Grants," *New Mexico Genealogical Society*, Albuquerque: New Mexico Genealogical Society, 1997, accessed March 22, 2018, http://pvhs.fms.k12.nm.us/teachers/mthur/086327A0-00757F35.6/New%20Mexico%20types%20of%20landgrants%20.pdf. Also see Hamilton, *Deeply Rooted*, for the story of Virgil Trujillo, "a tenth-generation rancher in New Mexico struggling to restore agriculture as a pillar of his community."

25. John R. Van Ness, "Hispanic Land Grants: Ecology and Subsistence in the Uplands of Northern NM and Southern Colorado," in Briggs and Van Ness, eds., *Land, Water, and Culture*, 186–87.

26. "Transcript of Treaty of Guadalupe Hidalgo (1848)," The Avalon Project at Yale Law School, OurDocuments.gov, National History Day, National Archives and Records Administration, and USA Freedom Corps, accessed January 12, 2012, http://www.ourdocuments.gov/doc.php?doc=26&page=transcript. Also see Weber, *Spanish Frontier in North America*; and Dr. L. M. Garcia y Griego, "The Aftermath of the Treaty of Guadalupe Hidalgo: Land Adjudication, Citizenship and Immigration," The Treaty of Guadalupe Hidalgo, A Living Document Symposium, Arizona State Museum and the University of Arizona, February 12, 2011, accessed March 22, 2018, https://www.youtube.com/watch?v=oTGZXXv4ptY.

27. Malcolm Ebright, "New Mexican Land Grants: The Legal Background," in Briggs and Van Ness, eds., *Land, Water, and Culture*, 25–26. For specific cases of land grant adjudication, see Malcolm Ebright, *Land Grants and Lawsuits in Northern New Mexico*. To address the differences between the treatment of Hispanic and Native American lands, see Christine A. Klein, "Treaties of Conquest: Property Rights, Indian Treaties, and the Treaty of Guadalupe Hidalgo," 201–55.

28. Knowlton, "Causes of Land Loss Among the Spanish Americans," 2.

29. Dunbar-Ortiz, *Roots of Resistance*, 7–8.

30. For more information on the subject, see Julian Haynes Steward, "The Concept and Method of Cultural Ecology," in Haenn and Wilk, eds., *The Environment in Anthropology*, 5–9. For the impact of ranching on cultural ecologies, see Arnold Strickon, "The Euro-American Ranching Complex," in Leeds and Vayda, eds., *Man, Culture, and Animals*, 229–58.

31. "Impact Los Alamos Project," Rubén Montoya, interview by Carlos Vásquez, Santa Fe, NM, August 9, 1994, Box 2, CDs 11–13.

32. Simmons, *Albuquerque, A Narrative History*, xv. On the toponymy of places in New Mexico, see Julyan, *The Place Names of New Mexico*.

33. Wood, "The Transformation of Albuquerque 1945–1972," 7, 51. Also see Stephen M. Wheeler and Wade Patterson, "The Rise of the Regional City:

Spatial Development of the Albuquerque Metropolitan Area," *New Mexico Historical Review* 82, no. 1 (Winter 2007).

34. Lazzell and Payne, *Historic Albuquerque*, 49.

35. Ibid., 51.

36. Marta Weigle, "Alluring New Mexico: Engineered Enchantment, 1880–1941," in Weigle, Levine, and Stiver, eds., *Telling New Mexico*, 235.

37. Brian McDonald, David Boldt, and University of New Mexico, "The New Mexico Economy," 8–9.

38. Bittman, "Dependency and the Economy of New Mexico," 39; Ferenc M. Szasz, "New Mexico During the Second World War," in Weigle, Levine, and Stiver, eds., *Telling New Mexico*, 294.

39. Rothman, *On Rims and Ridges*, 20.

40. Dunbar-Ortiz, *Roots of Resistance*, 119.

41. Nash, *The American West Transformed*, 11.

42. Bittman, "Dependency and the Economy of New Mexico," 46.

43. Ibid., 47.

44. "Impact Los Alamos Project," Ramón Frésquez, interview by Kenneth Salazar, San Pedro, NM, March 14, 1995, Box 1, CDs 42–43.

45. "Impact Los Alamos Project," Nick Salazar, interview by Carlos Vásquez, San Juan Pueblo, NM, July 29, 1994, Box 2, CDs 27–28.

46. "Impact Los Alamos Project," Florida Martínez, interview by Carlos Vásquez, Chimayó, NM, October 27, 1991, Box 1, CDs 58–59.

47. "Impact Los Alamos Project," Joe G. Montoya, interview by Carlos Vásquez, Cundiyo, NM, July 29, 1994, Box 2, CDs 7–8.

48. "Impact Los Alamos Project," Nick Salazar.

49. "Impact Los Alamos Project," José Benito Montoya, interview by Steve Fox, Pojoaque, NM, August 8, 1994, Box 2, CDs 9–10.

50. Kay, *Chimayó Valley Traditions*, 65.

51. "Impact Los Alamos Project," Bernadette V. Córdova, interview by Peggy Coyne, Española, El Wache, NM, February 29, 1996, Box 1, CDs 33–34.

52. "Impact Los Alamos Project," Josefita Velarde, interview by Peter Malmgren, November 20, 1995, Box 2, CDs 40–43.

53. "Impact Los Alamos Project," Nelson Kevin Vigil, interview by Carlos Vásquez, November 3, 1991, Box 2, CD 44.

54. Guthrie, *Recognizing Heritage*.

55. Steeper, *Gatekeeper to Los Alamos*, 22.

56. See Richard C. Dillon, "Governor's Message: A Word to Tourists," in Weigle, Levine, and Stiver, eds., *Telling New Mexico*, 243.

57. Mabel Dodge Luhan, "Lawrence of New Mexico," and D. H. Lawrence,

"New Mexico," *New Mexico Magazine*, February 1936, 10–11. See Henry Shukman, "D. H. Lawrence's New Mexico: The Ghosts That Grip the Soul of Bohemian Taos," *The New York Times*, October 22, 2006, http://www.nytimes .com/2006/10/22/travel/22culture.html?pagewanted=all&_r=0.

58. The Aztec Ruins, Bandelier, Gila Cliff Dwellings, Salinas Pueblo Missions, El Morro, Capulin Volcano, and White Sands Monuments and Carlsbad Caverns Park.

59. Kenneth W. Karsmizki in Campbell, *Magnificent Failure*, 3.

60. Campbell, *Magnificent Failure*, 29.

61. Dr. L. M. Garcia y Griego, "The Aftermath of the Treaty of Guadalupe Hidalgo."

62. "Impact Los Alamos Project," Ramón Frésquez.

63. "Impact Los Alamos Project," Nelson Kevin Vigil; Kurt F. Anschuetz and Thomas Merlan, "More than a Scenic Mountain Landscape: Valles Caldera National Preserve Land Use History," General Technical Report RMRS-GTR-196 (Fort Collins, CO: United States Department of Agriculture, Forest Service, and Rocky Mountain Research Station, September 2007), 28–29, accessed March 22, 2018, https://www.fs.fed.us/rm/pubs/rmrs_gtr196. pdf. For further reading, see Craig Martin, *Valle Grande: A History of the Baca Location No. 1*.

64. New Mexico Museum of Art, "History: The Great Depression and World War II," *New Mexico Art Tells New Mexico History*, accessed January 9, 2014, http://online.nmartmuseum.org/nmhistory/people-places-and-politics/ the-great-depression/history-the-great-depression-and-world-war-ii.html.

65. Sarah Deutsch, "Labor, Land, and Protest since Statehood," in Weigle, Levine, and Stiver, eds., *Telling New Mexico: A New History*, 276.

66. "Farm and Ranch Folks Project," David McDonald, interview by Jane O'Cain and Beth Morgan, the McDonald Residence, NM, June 4, 1997, Oral History Program (Las Cruces, NM: New Mexico Farm and Ranch Heritage Museum, New Mexico Department of Cultural Affairs, Research and Collections), accessed February 15, 2014, http://www.nmfarmandranchmuseum .org/oralhistory/detail.php?interview=207%27.

67. "Impact Los Alamos Project," Hipólita and Delfido Fernández, interview by Troy Fernández (their grandson), Chimayó, NM, February 27, 1994, Box 1, CDs 36–37. See Melzer, *Coming of Age in the Great Depression*.

68. "Impact Los Alamos Project," Montoya.

69. Michael Welsh, "The Land of Extremes: The Economy of Modern New Mexico," in Etulain, ed., *Contemporary New Mexico, 1940–1990*, 67.

70. DeBuys, *Enchantment and Exploitation*, 210.

71. Rothman, *On Rims and Ridges*, 178.

72. Southwest Crossroads Spotlight, "The Pajarito Plateau and Los Alamos," *Cultures and Histories of the American Southwest* (Santa Fe, NM: SAR Press and School for Advanced Research, 2006), accessed March 7, 2014, http://southwest crossroads.org/record.php?num=891.

73. Dorothy Hoard, "Historic Transportation Routes on the Pajarito Plateau," LA-UR-06-3550 (Los Alamos, NM: Los Alamos National Laboratory Ecology Group, Environmental Stewardship Division, May 2006), 4–5, accessed April 19, 2015, http://permalink.lanl.gov/object/tr?what=info:lanl-repo/lareport /LA-UR-06-3550.

74. Rothman, *On Rims and Ridges*, 16.

75. Theodore Roosevelt, "Proclamation 603—Establishment of Jemez Forest Reserve, New Mexico," October 12, 1905, accessed January 11, 2015, http:// www.presidency.ucsb.edu/ws/index.php?pid=69631.

76. Machen, McGehee, and Hoard, *Homesteading on the Pajarito Plateau, 1887–1942*, i.

77. Ibid., 3.

78. Jim Sagel, "Los Alamos: He Lived on the Hill Before It Meant 'The Bomb,'" *Albuquerque Journal North*, February 1, 1986; Sue Tester, Los Alamos Bureau, "Homesteaders: The Pioneers of Hilltop City," Santa Fe, NM, March 27, 1977, Center for Southwest Research Vertical Files, University Libraries, University of New Mexico, Los Alamos, NM—Impact of Manhattan Project on Area Communities, Homesteaders on Pajarito Plateau, 1942.

79. Rothman, *On Rims and Ridges*, 96.

80. Machen, McGehee, and Hoard, *Homesteading on the Pajarito Plateau*, 106.

81. Wirth and Aldrich, *Los Alamos*, 3, 28.

82. Los Alamos Ranch School, *Los Alamos Ranch School*.

83. LARS alumni include the presidents of General Motors, Quaker Oats, Sears and Roebuck, the owner of the Chicago White Sox and Oakland Athletics Bill Veek, and authors such as Gore Vidal and William S. Burroughs. Hunner, *Inventing Los Alamos*, 16.

84. Ibid., 36–37; Gonzales, *A Boy on the Hill*, 3.

85. Wirth and Aldrich, *The Ranch School Years*, 189.

86. Gonzales, *A Boy on the Hill*, 4.

87. Ibid., 26–27.

88. Church, *The House at Otowi Bridge*, 5.

89. Ibid.

90. Ibid., 25.

91. Ibid., 111.

92. Vincent B. Price, "Edith Warner in the Shadow of Los Alamos," *Provincial*

Matters, New Mexico Mercury, August 19, 2013, accessed January 30, 2014, http://newmexicomercury.com/blog/comments/provincial_matters_8_20 _2013.

Chapter Three

1. Hevly and Findlay, *The Atomic West,* 4.

2. Mason Sutherland and Justin Locke, "Adobe New Mexico," *The National Geographic Magazine,* December 1949, 826; Ralph Carlisle Smith Papers on Los Alamos 1924–1957, Albuquerque, NM: Center for Southwest Research, University Libraries, University of New Mexico, Collection MSS149BC, Box 1, Folder 62.

3. "One of Our 50 Is Missing," *New Mexico Magazine,* June 2014, accessed March 22, 2018, http://www.nmmagazine.com/article/?aid=86278#.V7yJaq 3LmdU. Also see Richard C. Sandoval, "One of Our Fifty is Missing: A Continuing Saga," *New Mexico Magazine,* 1986.

4. Montoya, *Translating Property,* 8, 9.

5. Masur, "Bernard DeVoto," 443.

6. See Barrillot, *Les irradiés de la République;* and Douglas Holdstock and Franck Barnaby, eds., *The British Nuclear Weapons Programme, 1952–2002.* See also Nic Maclellan, *Grappling with the Bomb* (Canberra: Australian University Press, 2017).

7. Masur, "Bernard DeVoto."

8. Frederick Jackson Turner wrote in his 1893 essay "The Significance of the Frontier in American History" that the disappearance of the Frontier marked the end of the first period in American History. His thesis has been debated and contested countless times by historians of the American West, but it had, nonetheless, a significant impact on the perception of the region in popular history as "free land" or, in other terms, as "empty" land. See Richard White, "Frederick Jackson Turner and Buffalo Bill," in *The Frontier in American Culture,* ed. James R. Grossman (Berkeley: University of California Press, 1994), 7–65, for an analysis of how Turner's and Buffalo Bill's contrasting accounts of the Frontier and its closing "came to be so culturally dominant and persistent." White writes that "Turner often placed himself and his audience not in the West but in popular representations of the West" (13).

9. Truman, "Statement by the President of the United States."

10. Jon Hunner uses the term "rejuvenate" to refer to "the frontier of science" that "preserved the tradition of the western part of the United States as a place of invention and vitality, of path-breaking innovations and pace-setting projects." Hunner, *J. Robert Oppenheimer,* 229.

11. Hart S. Horn, "Bonanzas and Buzz Bombs," *New Mexico Magazine*, December 1945, 10.

12. Max Coan, "Exile from Enchantment," *New Mexico Magazine*, March 1947, 21.

13. Hevly and Findlay explain that they borrowed the term from Patricia N. Limerick to use as the title of their work. See Limerick, "The Significance of Hanford in American History," in Hirt, ed., *Terra Pacifica*, 53–54.

14. Limerick, *Something in the Soil*, 24–25, 19, 2.

15. White, "It's Your Misfortune and None of My Own," 461–62.

16. H. R. Rodgers, "Land Provided for Bombing Range, Resources Unlimited, New Mexico Aids National Defense, A Department of Information and Reports of the State Land Office," *New Mexico Magazine*, June 1941, 31.

17. Michael Welsh, "The Land of Extremes: The Economy of Modern New Mexico," in Etulain, ed., *Contemporary New Mexico, 1940–1990*, 70.

18. Seth and Montgomery, Attorneys and Counselors at Law, Letter to Governor John Dempsey, Santa Fe, NM, December 28, 1942, Governor John Dempsey Papers, 1936–1958, Santa Fe: New Mexico State Records Center and Archive, Collection 1959–046, Box 13144, Folder 6, World War II Bases New Mexico Military.

19. The Farmers & Stockmens Bank, Letter to Governor John Dempsey, Clayton, NM, December 30, 1942, Governor John Dempsey Papers, 1936–1958, Santa Fe: New Mexico State Records Center and Archive, Collection 1959–046, Box 13144, Folder 6, World War II Bases New Mexico Military.

20. Letter from the Melrose Rotary Club to Governor Clinton P. Anderson, October 9, 1945, Harry S. Truman Presidential Library, Clinton P. Anderson Papers, Folder M.

21. Alberts and Putnam, *A History of Kirtland Air Force Base, 1928–1982*, 1.

22. Rodgers, "Land Provided for Bombing Range," 31.

23. Alberts et al., *A History of Kirtland Air Force Base*, 26.

24. "Holloman Air Force Base History," *Holloman Air Force Base*, United States Air Force, July 25, 2008, accessed March 31, 2014, http://www.holloman.af.mil/library/factsheets/factsheet.asp?id=4361.

25. "How It all Began," *White Sands Missile Range New Mexico* (San Diego, CA: MARCOA, 2007/2008), 8–9.

26. H. R. Rodgers, "Bombing Range Negotiations Progress, Resources Unlimited, New Mexico Aids National Defense, A Department of Information and Reports of the State Land Office," *New Mexico Magazine*, January 1942, 39.

27. Ibid.

28. Michael Welsh, *Dunes and Dreams: A History of White Sands National*

Monument, Professional Paper no. 55 (Santa Fe, NM: National Park Service, Division of History, Intermountain Cultural Resources Center, 1995), accessed April 19, 2015, http://www.nps.gov/whsa/learn/historyculture/upload/Dunes-and-Dreams.pdf.

29. See Roberts and Roberts, *New Mexico*, 166.

30. Rodgers, "Land Provided for Bombing Range, Resources Unlimited, New Mexico Aids National Defense, A Department of Information and Reports of the State Land Office," *New Mexico Magazine*, June 1941, 31.

31. Henry L. Stimson, Secretary of War Letter to Governor John Dempsey, Washington, DC, War Department, January 8, 1943, Governor John Dempsey Papers, 1936–1958, Santa Fe, NM: New Mexico State Records Center and Archive, Collection 1959–046, Box 13144, Folder 6, World War II Bases New Mexico Military. Emphasis added.

32. Lease and Suspension Agreement, War Department, Office of the Chief of Engineers, Construction Division, Real Estate Branch, CR-Form 143, September 24, 1942, Governor John Dempsey Papers, 1936–1958, Santa Fe: New Mexico State Records Center and Archive, Collection 1959–046, Box 13144, Folder 6, World War II Bases New Mexico Military.

33. Governor John Dempsey, Letter to Secretary of War Henry L. Stimson, January 27, 1943, Governor John Dempsey Papers, 1936–1958, Santa Fe: New Mexico State Records Center and Archive, Collection 1959–046, Box 13144, Folder 6, World War II Bases New Mexico Military.

34. Ibid.

35. Lewis N. Gillis, President of the Alamogordo Chamber of Commerce, "Proposed Bombing Range in Southwestern Part of Otero County, New Mexico," Letter to the El Paso Chamber of Commerce, Alamogordo, NM, August 18, 1943, Governor John Dempsey Papers, 1936–1958, Santa Fe: New Mexico State Records Center and Archive, Collection 1959–046, Box 13144, Folder 6, World War II Bases New Mexico Military.

36. Chris P. Fox, Executive Vice-President & General Manager of the El Paso Chamber of Commerce, Letter to Lewis N. Gillis, President of the Alamogordo Chamber of Commerce, August 20, 1943, Governor John Dempsey Papers, 1936–1958, Santa Fe: New Mexico State Records Center and Archive, Collection 1959–046, Box 13144, Folder 6, World War II Bases New Mexico Military.

37. Rodgers, "Bombing Range Negotiations Progress," 39.

38. "Acquire Bombing Range July First," *New Mexico Magazine*, May 1942, 38.

39. H. R. Rodgers, "Will Compensate Ranchers on Bombing Range, Resources Unlimited, New Mexico Aids National Defense, A Department of

Information and Reports of the State Land Office," *New Mexico Magazine*, June 1942, 29.

40. Rodgers, "Bombing Range Negotiations Progress," 39.

41. J. H. McLaughlin, Chairman of the Dona Ana County Commissioners, Western Union telegram to Governor John Dempsey, Hatch, NM, April 21, 1945, Governor John Dempsey Papers, 1936–1958, Santa Fe: New Mexico State Records Center and Archive, Collection 1959–046, Box 13144, Folder 6, World War II Bases New Mexico Military.

42. White Sands Missile Range Museum, White Sands Missile Range Ranching Heritage Oral History Project exhibit, visited on October 26, 2013. The Oral History Collection was prepared for White Sands Missile Range by Peter L. Eidenbach, Beth Morgan, and Mark Carter, eds., "Homes on the Range: Oral Recollections of Early Ranch Life on the U.S. Army White Sands Missile Range, New Mexico," US Department of Defense, Legacy Resource Management Program, Ranching Heritage Oral History Project, Las Cruces, NM: Human Systems Research, 1994. Emphasis added.

43. The information on the ranching era does not solely come from the museum exhibit but also from Thomas Widner et al., "Draft Final Report of the Los Alamos Historical Document Retrieval and Assessment (LAHDRA) Project" (Atlanta: Centers for Disease Control and Prevention, National Center for Environmental Health Division of Environment Hazards and Health Effects, Radiation Studies Branch, June 2009), accessed April 19, 2015, http://www.lahdra.org/pubs/Final%20LAHDRA%20Report%202010.pdf, chapter 10, 10–42, 10–43.

44. Eidenbach, Morgan, and Carter, "Homes on the Range."

45. Fritz Thompson, "Locals Witnessed History in a Flash," *Albuquerque Journal*, July 1995, accessed May 20, 2014, http://www.abqjournal.com/trinity/trinity2.htm.

46. "Farm and Ranch Folks Project," David McDonald, interview by Jane O'Cain and Beth Morgan, the McDonald Residence, NM, June 4, 1997.

47. "Farm and Ranch Folks Project," Irving Porter, interview by Donna M. Wojcik, the Porter Residence, Piñon, NM, October 9, 2009, Oral History Program, Las Cruces: New Mexico Farm and Ranch Heritage Museum, New Mexico Department of Cultural Affairs, Research and collections, accessed April 1, 2014, http://www.nmfarmandranchmuseum.org/oralhistory/detail.php?interview=226%27.

48. Rodgers, "Bombing Range Negotiations Progress," 39.

49. "Farm and Ranch Folks Project," Ernest Aguayo, Interview by Beth Morgan, Las Cruces, NM, June 14, 19, 25, 26 and July 11, 2001, Oral History Program, Las Cruces: New Mexico Farm and Ranch Heritage Museum, New Mexico Department of Cultural Affairs, Research and Collections, accessed

March 30, 2014, http://www.nmfarmandranchmuseum.org/oralhistory/detail.php?interview=15.

50. Before the launch of the Project, American atomic research was embryonic. It was a letter from the renowned German-born Nobel Laureate Albert Einstein to President Franklin D. Roosevelt (1933–1945) that led to allocating funds for the US atomic program. See Albert Einstein, Letter to President Roosevelt, August 2, 1939 (Document A2), in Stoff, Fanton, and Williams, eds., *The Manhattan Project*, 18–19.

51. Conant, *109 East Palace*, 37.

52. Groves, *Now It Can Be Told*, 64.

53. Hales, *Atomic Spaces*, 13.

54. John H. Dudley, "Ranch School to Secret City," in Badash, Hirschfelder, and Broida, eds., *Reminiscences of Los Alamos, 1943–1945*, 1–11.

55. Jones, *Manhattan, the Army, and the Atomic Bomb*, 83.

56. Wirth and Aldrich, *Los Alamos: The Ranch School Years*, 155.

57. Chambers, "Technically Sweet Los Alamos," 52.

58. Groves, *Now It Can Be Told*, 65, 289.

59. Edwin M. McMillan, "Early days at Los Alamos," in Badash, Hirschfelder, and Broida, eds., *Reminiscences of Los Alamos, 1943–1945*, 15.

60. Simmons, *Albuquerque: A Narrative History*, 367.

61. Groves, *Now It Can Be Told*, 66, 155.

62. Bird and Sherwin, *American Prometheus*, 81.

63. J. Robert Oppenheimer, "Crossing."

64. Wirth and Aldrich, *Los Alamos*, viii.

65. Bird and Sherwin, *American Prometheus*, 437.

66. Masco, *The Nuclear Borderlands*, 166.

Chapter Four

1. Hunner, *Inventing Los Alamos*, 15.

2. David Joshua Anderson, "Los Alamos Ranch and the Manhattan Project," NewMexicoHistory.org (Santa Fe, NM: The Office of the State Historian), accessed April 10, 2014, http://newmexicohistory.org/people/los-alamos-ranch-and-the-manhattan-project#_edn19.

3. Groves, *Now It Can Be Told*, 67.

4. Wirth and Aldrich, *Los Alamos*, 156–57.

5. Henry L. Stimson, Secretary of War, "Los Alamos Ranch School Seizure Letter," December 1, 1942. Available at "Civilian Displacement: Los Alamos, NM," *Atomic Heritage Foundation*, accessed March 22, 2018, https://www.atomicheritage.org/history/civilian-displacement-los-alamos-nm.

6. Kai Bird and Martin Sherwin, "A Los Alamos Beginning," in Kelly, ed., *Remembering the Manhattan Project*, 55.

7. Warner, *In the Shadow of Los Alamos*, 20.

8. Wirth and Aldrich, *Los Alamos*, 158.

9. Bartimus and McCartney, *Trinity's Children*, 79, 89.

10. McMillan, "Early Days at Los Alamos," in Badash, Hirschfelder, and Broida, eds., *Reminiscences of Los Alamos, 1943–1945*, 17–18.

11. Daniel Lang, "A Reporter in New Mexico," *The New Yorker*, April 17, 1948, 68, Ralph Carlisle Smith Papers on Los Alamos 1924–1957, Albuquerque: Center for Southwest Research, University Libraries, University of New Mexico, Collection MSS149BC, Box 1, Folder 24. Emphasis added.

12. Sagel, "Los Alamos: He Lived on the Hill Before It Meant 'The Bomb,'" *Albuquerque Journal North*, February 1, 1986.

13. Hales, *Atomic Spaces*, 59.

14. Diana Heil, "Justice Draws Near for Heirs of Land Taken by U.S. Government," *Santa Fe New Mexican*, October 12, 2004, B1, Fray Angélico Chávez History Library Vertical Files, Santa Fe, NM: Palace of the Governors, Department of Cultural Affairs, Los Alamos National Laboratories—Pajarito Plateau Hom 1.

15. Chuck Moutain, "Los Alamos Fire May Bring Heat and Justice to Original Latino Families," *Imagen Magazine*, September 2000, 26, Center for Southwest Research Vertical Files, University Libraries, University of New Mexico, Los Alamos—Health Hazards.

16. Malcolm Ebright, "Hispanic Homesteaders on the Pajarito Plateau: An Unconstitutional Taking of Property at Los Alamos, 1942–1945," *La Jicarita News*, Chamisal, NM: Rio Pueblo/Rio Embudo Watershed Protection Coalition, May 2007, 4, Albuquerque: Center for Southwest Research, University Libraries, University of New Mexico, Vertical Files, Los Alamos, NM—Impact of Manhattan Project on Area Communities, Homesteaders on Pajarito Plateau, 1942.

17. Sagel, "Los Alamos: He Lived on the Hill Before It Meant 'The Bomb.'"

18. Ibid., 5.

19. "Los Alamos, NM (Town Center)," *Neighborhood Scout*, Location Inc., accessed August 25, 2016, https://www.neighborhoodscout.com/nm/los-alamos/town-center/.

20. Sagel, "Los Alamos: He Lived on the Hill Before It Meant 'The Bomb.'"

21. The expression comes from one of President Franklin D. Roosevelt's most famous speeches, "The Great Arsenal of Democracy," delivered via radio on December 29, 1940. Building an atomic bomb was part of this arsenal. To read or listen to the whole speech, see Franklin D. Roosevelt, "The Great Arsenal of Democracy," December 29, 1940, *American Rhetoric*, accessed February 11, 2015, http://www.americanrhetoric.com/speeches/fdrarsenalofdemocracy.html.

22. Groves, *Now It Can Be Told*, 65.

23. Wirth and Aldrich, *Los Alamos*, 196.

24. Jones, *Manhattan, the Army, and the Atomic Bomb*, 86.

25. Chambers, "Technically Sweet," 1.

26. White, "It's Your Misfortune and None of My Own," 537.

27. "Authentic Albuquerque: Western Legends," *Albuquerque Convention and Visitors Bureau*, accessed April 30, 2014, http://www.visitalbuquerque.org/albuquerque/history/western/.

28. "Ghost Towns," *New Mexico True*, Santa Fe, NM: New Mexico Tourism Department, accessed April 30, 2014, http://www.newmexico.org/nm-adventures-ghost-towns/.

29. Hunner, *J. Robert Oppenheimer*, 79.

30. Fisher, *Los Alamos Experience*, 59.

31. Slotkin, *Regeneration Through Violence*, 7.

32. Ruth Marshak, "Secret City," in Serber and Wilson, eds., *Standing by and Making Do*, 2. Ruth Marshak was the wife of Robert Marshak, deputy group leader of the Theoretical Physics Division.

33. Jean Bacher, "Fresh Air and Alcohol," in Serber and Wilson. eds., *Standing by and Making Do*, 109–110.

34. Manley, "A New Laboratory is Born," Badash, Hirschfelder, and Broida, eds., *Reminiscences of Los Alamos, 1943–1945*, 29.

35. Fern Lyon, "The Atomic Age: Research Labs Keep New Mexico in the Forefront of Technology," *New Mexico Magazine*, October 1987, 60.

36. Kathleen Mark, "A Roof Over Our Heads," in Serber and Wilson, eds., *Standing by and Making Do*, 36.

37. Gibson and Michnovicz, *Los Alamos 1944–1947*, 81.

38. Charlie Masters, "Going Native," in Serber and Wilson, eds., *Standing by and Making Do*, 122.

39. Gibson and Michnovicz, *Los Alamos 1944–1947*, 103.

40. Jette, *Inside Box 1663*, 42.

41. Brode, *Tales of Los Alamos: Life on the Mesa, 1943–1945*, 52.

42. Marshak, "Secret City," 3.

43. Elsie McMillan, "Outside the Inner Fence," in Badash, Hirschfelder, and Broida, eds., *Reminiscences of Los Alamos, 1943–1945*, 41. Emphasis added.

44. Rhodes, *The Making of the Atomic Bomb*, 540. Emphasis added.

45. Fisher, *Los Alamos Experience*, 44. Emphasis added.

46. Hales, *Atomic Spaces*, 207.

47. Berenice Brode, "Tales of Los Alamos," in *Reminiscences of Los Alamos, 1943–1945*, 157. Emphasis added.

48. Ibid., 154.

49. Masters, "Going Native," 122.

50. Brode, *Tales of Los Alamos*, 52.

51. Masters, "Going Native," 124, 128–29.

52. Brode, "Tales of Los Alamos, 153.

53. Fisher, *Los Alamos Experience*, 89.

54. Brode, "Tales of Los Alamos," 154.

55. Brode, *Tales of Los Alamos*, 84, 138.

56. Fisher, *Los Alamos Experience*, 36, 87, 89.

57. Mason, *Children of Los Alamos*, xi.

58. Interview of Martha Bacher Eaton in Mason, *Children of Los Alamos*, 54.

59. Church, *The House at Otowi Bridge*, 107.

60. Interview of Jim Bradbury in Mason, *Children of Los Alamos*, 121.

61. Interview of David Bradbury in Mason, *Children of Los Alamos*, 157.

62. Interview of Severo Gonzales in Mason, *Children of Los Alamos*, 58–59.

63. Interview of Dimas Chavez in Mason, *Children of Los Alamos*, 165.

64. Dimas Chavez, interview by Cindy Kelly, Washington, DC, February 13, 2013, Atomic Heritage Foundation and Los Alamos Historical Society, "Voices of the Manhattan Project," Washington, DC: Atomic Heritage Foundation, Los Alamos, NM: Los Alamos Historical Society, 2012, accessed May 2, 2014, http://www.manhattanprojectvoices.org/oral-histories/dimas-chavezs-interview.

65. "Impact Los Alamos Project," Bernadette V. Córdova.

66. "Impact Los Alamos Project," José Benito Montoya.

67. "Impact Los Alamos Project," Joe G. Montoya.

68. "Impact Los Alamos Project," Richard Cook, interview by Steve Fox, Española, NM, August 8, 1995, Box 1, CD 31.

69. "Impact Los Alamos Project," Delfido Fernández.

70. "Impact Los Alamos Project," Sennu A. Gallegos, interview by Carlos Vásquez, LANL, NM, March 3, 1991, Box 1, CDs 45–46.

71. "Impact Los Alamos Project," Genaro Martínez.

72. "Impact Los Alamos Project," Carlos Vásquez, University of New Mexico Conference Center, Albuquerque, March 7, 1996, Center for Southwest Research, University Libraries, University of New Mexico, MSS 821 Box 2 CD50.

73. Carlos Vásquez, "Impact Los Alamos: Traditional New Mexico in a High Tech World, Overview of Project and Symposia," *New Mexico Historical Review* 72 (January 1997): 7.

74. Darryl Martínez, interview by the author, Governor's Office of San Ilde-fonso Pueblo, NM, September 17, 2013.

75. Interview of Nella Fermi Weiner in Mason, *Children of Los Alamos*, 78.

76. Masco, *The Nuclear Borderlands*, 123.

Chapter Five

1. Groves, *Now It Can Be Told*, 289.

2. US Department of the Interior, Annual Report of the Secretary of the Interior for the Fiscal Year Ended June 30, 1938, 210.

3. Hacker, *The Dragon's Tail*, 85.

4. Larry Calloway, "N. M. Gave Birth to Atomic Bomb," *Albuquerque Journal*, September 19, 1999, Ferenc M. Szasz Papers, 1894–2005, Albuquerque: Center for Southwest Research, University Libraries, University of New Mexico, Collection MSS552BC, Box 9, Folder 11, Trinity Site Recollections, 1945–1999.

5. Gregory Walker, "Trinity Site," *Trinity Atomic Web Site*, 1995–2005, accessed May 16, 2014, http://www.abomb1.org/trinity/trinity1.html.

6. DeGroot, *The Bomb*, 56.

7. Simmons, *Albuquerque*, 367.

8. Hales, *Atomic Spaces*, 307.

9. Hacker, *The Dragon's Tail*, 75.

10. Clarfield and Wiecek, *Nuclear America*, 51.

11. Kunetka, *City of Fire*, 170.

12. Gregory Walker, "Trinity Atomic Test July 16, 1945," *Trinity Atomic Web Site*, 1995–2005, accessed March 22, 2018, http://www.abomb1.org/trinity/tr_test.html#Quotes.

13. Coan, "Exile from Enchantment," 21.

14. Dennis Eskow (science ed.), "How They Hid the Bomb," *Popular Mechanics*, August 1985, Ferenc M. Szasz Papers, 1894–2005, Albuquerque: Center for Southwest Research, University Libraries, University of New Mexico, Collection MSS552BC, Box 9, Folder 11, Trinity Site Recollections, 1945–1999.

15. Thompson, "Locals Witnessed History in a Flash."

16. Elvis E. Fleming, "Civilian Reaction To The First Atomic Bomb Test," Address at the Roswell Public Library, July 17, 1983, Ferenc M. Szasz Papers, 1894–2005, Albuquerque: Center for Southwest Research, University Libraries, University of New Mexico, Collection MSS552BC, Box 8, Folder 18, Elvis E. Fleming—"Local Reaction to the First Atomic Bomb."

17. Cantelon, Hewlett, and Williams, eds., *The American Atom*, 54.

18. Rolf Sinclair, "The Blind Girl Who Saw the Flash of the First Nuclear Weapon Test," *Skeptical Inquirer* 18 (Fall 1993): 63–67, Ferenc M. Szasz Papers, 1894–2005, Albuquerque: Center for Southwest Research, University Libraries, University of New Mexico, Collection MSS552BC, Box 9, Folder 11, Trinity Site Recollections, 1945–1999, and in Groves, *Now It Can Be Told*, 435.

19. David Smollar, "First A-Test Site: Bleak Desert Spot," *Los Angeles Times*,

July 16, 1985, Ferenc M. Szasz Papers, 1894–2005, Albuquerque: Center for Southwest Research, University Libraries, University of New Mexico, Collection MSS552BC, Box 8, Folder 7, Trinity 40th Anniversary, 1985.

20. All the quotes and testimonies are from Thompson, "Locals Witnessed History in a Flash"; Calloway, "N. M. Gave Birth to Atomic Bomb"; and Fleming, "Civilian Reaction to the First Atomic Bomb Test" in Ferenc M. Szasz Paper Collection.

21. Ferenc M. Szasz and George E. Webb, "New Mexican Response to the End of the Second World War," *New Mexico Historical Review* 83, no. 1 (Winter 2008): 20.

22. Szasz, *Larger Than Life*, 127, 130.

23. Hacker, *The Dragon's Tail*, 84–85.

24. Widner et al., *Draft Final Report of the LAHDRA Project*, chaps. 10, 26.

25. Ibid., chaps. 10, 22.

26. Caufield, *Multiple Exposures*, 58.

27. The measuring unit has now become the rem (roentgen equivalent man). The rem and the roentgen are comparable since one roentgen deposits about 0.96 rem in biological tissue. Since 1957, 5,000 millirems per year has been the occupational limit for an adult worker, including soldiers who are exposed to radiation. In addition, lifetime cumulative exposure should not exceed the equivalent of the person's age multiplied by 1,000 millirems. Information from the Massachusetts Institute of Technology, "Radiation, How Much Is Considered Safe for Humans?," *MIT News*, accessed May 19, 2014, http://newsoffice. mit.edu/1994/safe-0105.

28. Joseph O. Hirschfelder, "The Scientific and Technological Miracle at Los Alamos," in Badash, Hirschfelder, and Broida, eds., *Reminiscences of Los Alamos, 1943–1945*, 74. Also see J. O. Hirschfelder and John Magee, "Danger from Active Material Falling from Cloud—Desirability of Bonding Soil Near Zero with Concrete and Oil," and "Improbability of Danger from Active Material Falling from Cloud," Memos to K. T. Bainbridge, LA1027DEL, Los Alamos, NM, June 16, 1945 and July 6, 1945, LA1027DEL, Ferenc M. Szasz Papers, 1894–2005, Albuquerque: Center for Southwest Research, University Libraries, University of New Mexico, Collection MSS552BC, Box 8, Folder 47; Memos by J. O. Hirschfelder and John Magee Concerning Fate of the Active Material Following the Trinity Shot, Document LA 1027, 1945.

29. Hales, *Atomic Spaces*, 304.

30. Caufield, *Multiple Exposures*, 58.

31. Hirschfelder, "The Scientific and Technological Miracle at Los Alamos," 76–77.

32. Widner et al., *Draft Final Report of the LAHDRA Project*, chaps. 10, 37.

Historian Catherine Caufield also points out that "worried about lawsuits, Robert Oppenheimer . . . ordered the Health Group's Trinity reports to be held in the strictest secrecy. They were kept separate from other Trinity reports and could be released only with Oppenheimer's personal approval." *Multiple Exposures*, 61.

33. Hales, *Atomic Spaces*, 328.

34. Hacker, *The Dragon's Tail*, 104–5.

35. Widner et al., *Draft Final Report of the LAHDRA Project*, chaps. 10, 37.

36. Ibid. Also see Thomas E. Widner, "The World's First Atomic Blast and How It Interacted with the Jornada del Muerto and Chupadera Mesa," Geology of the Chupadera Mesa Region, 60th Field Conference, *New Mexico Geological Society Guidebook*, 2009, 425–28.

37. Ty Bannerman, "The First," *The American Literary Review*, Denton: University of North Texas, Department of English, Creative Writing Program, accessed April 21, 2015, http://www.americanliteraryreview.com/ty-bannerman-the-first.html.

38. Joan E. Price, "Downwinders Speak Out," *Desert Exposure*, OPC News, August 2015, accessed April 21, 2016, http://www.desertexposure.com/2015 08/201508_downwinders.php.

39. Wright H. Langham, Biomedical Research Group Leader, Letter to Madame Jacqueline Juillard, Ingénieur-Chimiste EPUL-SIA, Colovrex, Geneva, Switzerland, January 11, 1960, Ferenc M. Szasz Papers, 1894–2005, Albuquerque: Center for Southwest Research, University Libraries, University of New Mexico, Collection MSS552BC, Box 8, Folder 45, Livestock and Radiation, Document LASL 431, and letter about atomic bomb effect on Alamogordo Cattle, 1948–1961.

40. See Rosenberg, *Atomic Soldiers*, and Joseph Masco, *The Nuclear Borderlands*, "Above-ground Testing (1945–1962): Tactility and the Nuclear Sublime," 55–67.

41. Widner et al., *Draft Final Report of the LAHDRA Project*, chaps. 10, 27.

42. Ibid., chaps. 10, 40.

43. Ibid., chaps. 10, 37.

44. Mason, *Children of Los Alamos*, 54.

45. Szasz, *Larger than Life*, 128.

46. Bartimus and McCartney, *Trinity's Children*, 17.

47. Widner et al., *Draft Final Report of the LAHDRA Project*, chaps. 10, 41. For more information on the effects of Trinitite and the National Historic Site proceedings at Trinity, see Leslie P. Arnberger, Regional Director, Southwest Region, "Review of Proposals for Establishment of Trinity National Historic Site," Memo to Chief, Office of Park Planning and Environmental Agency, L58(SWR)PP, Santa Fe, NM, April 11, 1979, Ferenc M. Szasz Papers, 1894–

2005, Albuquerque: Center for Southwest Research, University Libraries, University of New Mexico, Collection MSS552BC, Box 8, Folder 48, Trinity Site 1945–1979.

48. Szasz, *Larger Than Life*, 128.

49. Glenn T. Seaborg, Chairman of the US Atomic Energy Commission, "Los Alamos: 25 Years in the Service of Science and the Nation," Remarks at the 25th Anniversary Celebration of the Los Alamos Scientific Laboratory, Washington, DC: US Atomic Energy Commission, February 15, 1968, Ferenc M. Szasz Papers, 1894–2005, Albuquerque: Center for Southwest Research, University Libraries, University of New Mexico, Collection MSS552BC, Box 7, Folder 7, US Atomic Energy Commission (AEC) and Los Alamos Scientific Laboratory, 1948–1968.

50. Hunner, *Inventing Los Alamos*, 98–99.

51. Szasz, *Larger Than Life*, 136.

52. See documents copied from the Los Alamos National Laboratory Records Center and Archives in University Libraries, University of New Mexico, Center for Southwest Research, MSS552BC, Ferenc M. Szasz Papers, 1894–2005, Box 7, Folder 4, Documents on Los Alamos, 1944–1946.

53. Norris Bradbury, "Los Alamos—the First 25 years," in Badash, Hirschfelder, and Broida, eds., *Reminiscences of Los Alamos, 1943–1945*, 163.

54. Linda Aldrich, "Zia Company 1946–1986," NewMexicoHistory.org, Santa Fe, NM: The Office of the State Historian, accessed May 28, 2014, http://newmexicohistory.org/people/zia-company-1946-1986.

55. See Ferguson, *Civilization: The West and the Rest*. Ferguson argues that "the West" conquered its rivals, partly, by its scientific superiority and that the rest of the world suffered from its lack of modern science and technology, which were, fundamentally, western products. See also Stephanson, *Manifest Destiny*. Stephanson argues that the process of Manifest Destiny can be traced back to the British settlement of northern America. He uses the Manifest Destiny expression to refer to a belief in a "providentially assigned role of the United States to lead the world to new and better things" (xii).

Chapter Six

1. US National Archives and Records Administration, "World War II Army Causalities: New Mexico," National Archives, accessed February 18, 2015, http://www.archives.gov/research/military/ww2/army-casualties/new-mexico.html.

2. Steeper, *Gatekeeper to Los Alamos*, 113.

3. Bird and Sherwin, *American Prometheus*, 422.

4. DeGroot, *The Bomb*, 176.

5. Hunner, *Inventing Los Alamos*, 163.

6. DeGroot, *The Bomb*, 159; "Budget," Los Alamos National Laboratory, Los Alamos, NM: Los Alamos National Security, LLC for the United States Department of Energy's National Nuclear Security Administration, accessed February 18, 2015, http://www.lanl.gov/about/facts-figures/budget.php.

7. Clarfield and Wiecek, *Nuclear America*, 89, 83.

8. DeGroot, *The Bomb*, 161.

9. George Fitzpatrick, "Los Alamos . . . the World's Most Important Small Town," *New Mexico Magazine*, August 1949, 21–22. Emphasis added.

10. George Fitzpatrick, "The Secret of Los Alamos," *New Mexico Magazine*, September 1945, 10, 11.

11. Fitzpatrick, "Los Alamos . . . the World's Most Important Small Town," 61.

12. "Impact Los Alamos Project," Maxine Beckman, interview by Linda Campbell, February 11, 1995, Box 1, CD 26.

13. "Impact Los Alamos Project," Rubén Waldo Salazar, interview by Peggy Coyne, Chamita, NM, February 21, 1996, Box 2, CD 29.

14. See Martin, *Quads, Shoeboxes, and Sunken Living Rooms*.

15. Paul M. Sears, and University of New Mexico Bureau of Business Research, "Los Alamos—Boom Town Under Control," *Business Backgrounds*, no. 1 (May 1953), Ralph Carlisle Smith Papers on Los Alamos 1924–1957, Albuquerque: Center for Southwest Research, University Libraries, University of New Mexico, Collection MSS149BC, Box 2, Folder 70.

16. Robert E. McKee, *The Zia Company in Los Alamos*, 17.

17. "Impact Los Alamos Project," Maxine Beckman.

18. "Impact Los Alamos Project," Harold Gibson, interview by Peter Malmgren, Chimayó, NM, February 12, 1996, Box 1, CDs 47–48.

19. See Stephanie B. Turner, "The Case of the Zia: Looking Beyond Trademark Law to Protect Sacred Symbols," *Yale Law School Legal Scholarship Repository*, Student Scholarship Papers, Paper 124, January 3, 2012, accessed February 15, 2016, http://digitalcommons.law.yale.edu/student_papers/124?utm_source=digitalcommons.law.yale.edu%2Fstudent_papers%2F124&utm_medium=PDF&utm_campaign=PDFCoverPages.

20. "Impact Los Alamos Project," Patricia Trujillo-Oviedo, interview by Carlos Vásquez, Chimayó, NM, October 27, 1991, Box 2, CDs 36–37.

21. All quotes are from McKee, *The Zia Company*, 7, 1, 9, 32–33, 20–21, and 23.

22. Mason Sutherland and Justin Locke, "Adobe New Mexico," *The National Geographic Magazine*, December 1949, 813.

23. Sears et al., "Los Alamos—Boom Town Under Control."

24. Darryl Martínez, interview by the author, Governor's Office of San Ildefonso Pueblo, NM, September 17, 2013.

25. Daniel Lang, "A Reporter in New Mexico," *The New Yorker*, April 17, 1948, 76, Ralph Carlisle Smith Papers on Los Alamos 1924–1957, Albuquerque: Center for Southwest Research, University Libraries, University of New Mexico, Collection MSS149BC, Box 1, Folder 24.

26. Margaret Page Hood, "No More Land," *New Mexico Magazine*, October 1945, 43.

27. "Impact Los Alamos Project," Aaron Martínez, interview by Peter Malmgren, February 24, 1996, Box 1, CDs 55–56.

28. David, ed., *Internalized Oppression*, 8.

29. Joe Alex Morris, "The Cities of America: Los Alamos," *The Saturday Evening Post*, December 11, 1948, Ralph Carlisle Smith Papers on Los Alamos 1924–1957, Albuquerque: Center for Southwest Research, University Libraries, University of New Mexico, Collection MSS149BC, Box 1, Folders 28, 29, 30.

30. "Impact Los Alamos Project," Rubén Waldo Salazar.

31. "Impact Los Alamos Project," Bernadette V. Córdova.

32. "Impact Los Alamos Project," Loyda Martínez, interview by Peter Malmgren, Chimayó, NM, December 3, 1995, Box 1, CDs 65–66.

33. "Impact Los Alamos Project," Nick Salazar, interview by Carlos Vásquez, San Juan Pueblo, NM, July 29, 1994. Salazar worked in Los Alamos until 1991, first in the procurement division and then as a technician.

34. "Impact Los Alamos Project," Hipólita Fernández.

35. "Impact Los Alamos Project," Alfonso Mascarenes, Interview by Dot Waldrip, 14 January 1996, Box 1, CDs 69–71.

36. "Impact Los Alamos Project," Pedro Martínez, interview by Peggy Coyne, February 15, 1996, Box 1, CDs 67–68.

37. "Impact Los Alamos Project," Lebeo Martínez.

38. "Impact Los Alamos Project," Danny Martínez, interview by Carlos Vásquez, LANL, NM, November 8, 1991, Box 1, CD 57.

39. "Impact Los Alamos Project," Pedro Martínez.

40. "Impact Los Alamos Project," C. L. Hunter, interview by Steve Fox, Española, August 9, 1994, Box 1, CDs 51–52.

41. "Impact Los Alamos Project," José Benito Montoya.

42. "Impact Los Alamos Project," Danny Martínez.

43. "Impact Los Alamos Project," Ed Sitzberger, interview by Linda Campbell, Cimarron, NM, February 10, 1995, Box 2, CDs 32–33.

44. "Impact Los Alamos Project," Lucille Sanchez, interview by Carlos Vásquez, October 27, 1991, Box 2, CDs 30–31.

45. "Impact Los Alamos Project," Rubén Montoya.

46. "Impact Los Alamos Project," Ramón Frésquez.

47. "Impact Los Alamos Project," Charles Montaño, interview by Carlos Vásquez, April 16, 1996, Box 2, CDs 1–4.

48. "Impact Los Alamos Project," Aaron Martínez.

49. Sears et al., "Los Alamos—Boom Town Under Control."

50. Ibid.

51. "Impact Los Alamos Project," Fransisco Leroy Pacheco, interview by Carlos Vásquez, Albuquerque, December 3, 1993, Box 2, CDs 20–24.

52. Furman, *Sandia National Laboratories*, 141, 151.

53. David A. Taylor, "Sandia Pioneers: Nuclear Bomb a Challenge for Young Scientists," *New Mexico* 83, no. 9 (September 2005): 60, Fray Angélico Chávez History Library Vertical Files, Santa Fe, NM: Palace of the Governors, Department of Cultural Affairs, Sandia National Laboratories 1. Furman, *Sandia National Laboratories*, 192.

54. Alexander, *History of Sandia Corporation*, 23; Furman, *Sandia National Laboratories*, 334.

55. Since 1993, the Sandia Corporation has been a subsidiary of the Lockheed Martin Corporation, the biggest defense contractor in the United States.

56. Ibid., 17.

57. Alexander, *History of Sandia Corporation*, 34.

58. Furman, *Sandia National Laboratories*, 224.

59. Aaron Baca, "Sandia Deal to Make Jobs," *Albuquerque Journal*, May 2, 2003, D-3, Fray Angélico Chávez History Library Vertical Files, Santa Fe, NM: Palace of the Governors, Department of Cultural Affairs, Sandia National Laboratories 1.

60. Furman, *Sandia National Laboratories*, 468.

61. Fremont Kutnewsky, "Research at Sandia," *New Mexico Magazine*, July 1958, 49.

62. Ibid., 72.

63. US Bureau of the Census, *U.S. Census of Population, 1950, Characteristics of Population*, Part 31, New Mexico, Washington, DC: Government Printing Office, 1952.

64. Wood, "The Transformation of Albuquerque," 190.

65. Karafantis, "Weapons Labs and City Growth," 99, 103, 118.

66. Ibid., 89, 98, 96.

67. Simmons, *Albuquerque*, 370.

68. Karafantis, "Weapons Labs and City Growth," 103.

69. Simmons, *Albuquerque*, 371.

70. Hart S. Horn, "Bonanzas and Buzz Bombs," *New Mexico Magazine*, December 1945.

71. George Fitzpatrick, "Alamogordo . . . City of the Rockets," *New Mexico Magazine*, January 1948, 17.

72. Twenty-one of these scientists, including Wernher von Braun, are profiled by Annie Jacobson in *Operation Paperclip*.

73. Orren Beaty, "Proving Ground for Rockets," *New Mexico Magazine*, March 1947, 54.

74. Fitzpatrick, "Alamogordo . . . City of the Rockets," 18.

75. George Fitzpatrick, "Hush-hush no.2," *New Mexico Magazine*, March 1954, 11–13.

76. Ibid., 47.

77. Joseph H. Forsyth, "They Track the Rocket Missiles," *New Mexico Magazine*, April 1955, 15.

78. Kathy Hedges, "A Brief History of New Mexico Tech," *New Mexico Tech*, Socorro, NM: New Mexico Institute of Mining and Technology, accessed June 14, 2014, http://www.nmt.edu/fast-facts/298-a-brief-history-of-nmt.

79. Wayne Winters, "Uranium Boom at Grants," *New Mexico Magazine*, March 1951, 14.

80. Ibid., 13.

81. Ibid., 14.

82. Toby Smith, "Grants—Boomtown!," *New Mexico Magazine*, September 1979, 54.

83. Kevin J. Fernlund, "Mining the Atom: The Cold War Comes to the Colorado Plateau, 1948–1958," *New Mexico Historical Review* 69, no. 4 (October 1994): 350.

84. Amundson, *Yellowcake Towns*, 79.

85. Bob Tucker, "Uranium Discovery Makes Laguna Indians Modern 'Rags to Riches' Story," July 14, 1957, Governor John Dempsey Papers, 1936–1958, Santa Fe, NM: New Mexico State Records Center and Archive, Collection 1959–046, Box 13191, G403, Folder 271, AEC Commission Subcommittee on Raw Materials Statements, Reports, 85th Congress 1957.

86. Amundson, *Yellowcake Towns*, 87, 88, 92, 93.

87. Szasz, *Larger Than Life*, 122.

Chapter Seven

1. Nash, *Federal Landscape*, 52–53. Nash bases much of his thesis on the development of California, but it is not comparable to most western states and certainly not to New Mexico, whose historical development within the United States was much slower and less autonomous than that of California. See Paul

Rhode, "The Nash Thesis Revisited." Rhode argues that California was already a pacesetter since it was already as large as, or even larger than, the rest of the West combined both in population and income in 1940.

2. See Fraser and Gerstle, eds., *The Rise and Fall of the New Deal Order 1930–1980*. Earl S. Pomeroy, the father of the "federal school," introduced a tradition in American history by focusing on the role of the federal government as a dominant force in reshaping the West as early as the nineteenth century. See Pomeroy, *The Territories and the United States*.

3. Howard R. Lamar, "Earl Pomeroy, Historian's Historian," *Pacific Historical Review* 56, no. 4 (November 1987): 546.

4. See Schumpeter, *The Theory of Economic Development*; also Saad, *Development Through Technology Transfer*.

5. Wheeler and Patterson, "The Rise of the Regional City," 6.

6. I borrow the expression from Edward H. Spicer, *Cycles of Conquest*.

7. White, "It's Your Misfortune and None of My Own," 461–62, 267.

8. Amundson, *Yellowcake Towns*, xvii, 175.

9. The covers are available at "The Uranium Rush–1949," *National Radiation Instrument Catalog*, accessed February 20, 2015, http://national-radiation -instrument-catalog.com/new_page_14.htm.

10. Amundson, *Yellowcake Towns*, 93.

11. Ibid., 108.

12. Virginia T. McLemore and New Mexico Bureau of Geology and Mineral and Resources, Socorro, NM, "Uranium Resources in New Mexico," Society of Mining, Metallurgy & Exploration Annual Meeting, Denver, CO, February 25–28, 2007.

13. Whan, chairman, "The State of New Mexico: Governor's Energy Task Force," 1.

14. Clark F. Huffman et al., "Petitioners v. Western Nuclear, Inc., et al. 486 U. 663 (108 S.Ct. 2087, 100 L.Ed.2d 693), No. 87–645," Cornell University Law School, Legal Information Institute, accessed August 18, 2014, http://www.law .cornell.edu/supremecourt/text/486/663.

15. Amundson, "Home on the Range No More," 498–99.

16. Huffman et al., *Petitioners v. Western Nuclear, Inc.*

17. The similarity with other boom-and-bust stories such as the decline of the automobile industry in Flint, Michigan, is striking. General Motors' policies in the 1980s and 1990s resulted in the laying off of close to 50,000 people. Like in Grants, tourism was one solution put forward by city officials as a way out of depression with the opening of an indoor theme park called "AutoWorld" (a commercial failure). With the rise of criminality in the

poverty-stricken neighborhoods, some of the employees laid off by GM found work as prison guards at the new county jail. This is documented in Michael Moore's 1989 documentary *Roger & Me* (Burbank, CA: Warner Brothers).

18. Power, "An Economic Evaluation of a Renewed Uranium Mining Boom in New Mexico," 1–2.

19. Eichstaedt, *If You Poison Us*, 163.

20. Alexander, *History of Sandia Corporation*, 36.

21. All quotes are from Krischner Associates, Management and Economic Consultants, *Adjustments to Reduced National Defense Expenditures in New Mexico*, 1, 2–4, 7, 10, 12.

22. McDonald, *The New Mexican Economy*, 18.

23. All quotes are from Krischner Associates, Management and Economic Consultants, *Adjustments to Reduced National Defense Expenditures in New Mexico*, 1, 2–4, 7, 10, 12.

24. "Atomic Research and Development," *New Mexico Magazine*, October 1960, 8, 11.

25. Dwight D. Eisenhower, "Atoms for Peace (8 December 1953)," Voices of Democracy: The U.S. Oratory Project, ed. Shawn J. Parry-Giles, accessed February 25, 2015, http://voicesofdemocracy.umd.edu/.

26. "Says Gnome to Hurt Area," Ferenc M. Szasz Papers, 1894–2005, Albuquerque: Center for Southwest Research, University Libraries, University of New Mexico, Collection MSS552BC, Box 14, Folder 36, Clippings on Project Gnome, 1953–1985. Also see Ferenc M. Szasz, "New Mexico's Forgotten Nuclear Tests: Projects Gnome (1961) and Gasbuggy (1967)," *New Mexico Historical Review* 3 (October 1998): 347–70.

27. Bryan Jeter, "Carlsbad 'Quietly Jubilant,'" Ferenc M. Szasz Papers, 1894–2005, Albuquerque: Center for Southwest Research, University Libraries, University of New Mexico, Collection MSS552BC, Box 14, Folder 36 Clippings on Project Gnome, 1953–1985.

28. Mike Hill, "Atom Bomb May Be Boon for Carlsbad," *Current-Argus*, January 25, 1959, Ferenc M. Szasz Papers, 1894–2005, Albuquerque: Center for Southwest Research, University Libraries, University of New Mexico, Collection MSS552BC, Box 14, Folder 36, Clippings on Project Gnome, 1953–1985.

29. "AEC Contracts with Mines to Pay for Lost Production," December 7, 1961, Ferenc M. Szasz Papers, 1894–2005, Albuquerque: Center for Southwest Research, University Libraries, University of New Mexico, Collection MSS552BC, Box 14, Folder 36, Clippings on Project Gnome, 1953–1985.

30. "Teller Calls Test Miracle of Ages," December 10, 1961, Ferenc M. Szasz

Papers, 1894–2005, Albuquerque: Center for Southwest Research, University Libraries, University of New Mexico, Collection MSS552BC, Box 14, Folder 36, Clippings on Project Gnome, 1953–1985.

31. Gerber, Hamburger, and Hull, *Plowshare*, 32.

32. US Atomic Energy Commission Press Release, Nevada Operation Office, NV-67-9, Las Vegas, NV, January 31, 1967, Ferenc M. Szasz Papers, 1894–2005, Albuquerque: Center for Southwest Research, University Libraries, University of New Mexico, Collection MSS552BC, Box 14, Folder 40, US Atomic Energy Commission Press Releases for Project Gasbuggy, 1967–1968.

33. Szasz, *Larger Than Life*, 170.

34. Welsh, "The Land of Extremes: The Economy of Modern New Mexico," in Etulain, ed., *Contemporary New Mexico, 1940–1990*, 77.

35. See Powaski, *March to Armageddon*, 72.

36. Wood, "The Transformation of Albuquerque," 534.

37. Simmons, *Albuquerque: A Narrative History*, 377.

38. Ibid., 374.

39. Wood, "The Transformation of Albuquerque," 297–98.

40. "Atomic Research and Development," *New Mexico Magazine*, October 1960, 8, 11.

41. Maria E. Montoya, "Dennis Chavez and the Making of Modern New Mexico," in Weigle, Levine, and Stiver, eds., *Telling New Mexico*, 342.

42. Domenici, *A Brighter Tomorrow*, 3.

43. Whan, "The State of New Mexico: Governor's Energy Task Force," 1–2.

44. New Mexico Energy Institute, "Attitudes of New Mexico Residents toward the Nuclear Fuel Industry," NMEI Report No. 76–513A, Albuquerque: University of New Mexico, January 1977, 2, 3, 7.

45. McDonald, *The New Mexican Economy*, 2.

46. Ian McCullough, "Why did the U.S. Government Shut Down in October 2013?," Forbes.com, March 10, 2013, accessed February 17, 2015, http://www.forbes.com/sites/quora/2013/10/03/why-did-the-u-s-government-shut-down-in-october-2013/.

47. Dan Boyd and T. S. Last, "Impact of Shutdown Wide-Ranging in New Mexico," *Albuquerque Journal*, October 2, 2013, accessed February 17, 2015, http://www.abqjournal.com/273084/news/impacts-in-new-mexico-wideranging.html.

48. Rob Nikolewski, "How to Make NM less Vulnerable to Federal Gov't Shutdowns," Watchdog.org, October 15, 2013, accessed February 17, 2015, http://watchdog.org/19608/nm-how-to-make-nm-less-vulnerable-to-federal-govt-shutdowns-2/.

Chapter Eight

1. J. Robert Oppenheimer, "Speech to the Association of Los Alamos Scientists," November 2, 1945, Los Alamos, NM, Atomicarchive.com, accessed August 31, 2014, http://www.atomicarchive.com/Docs/ManhattanProject/OppyFarwe ll.shtml.

2. Hunner, *J. Robert Oppenheimer*, 205.

3. Hales, *Atomic Spaces*, 246.

4. Ferenc M. Szaz and George E. Webb, "New Mexican Response to the End of the Second World War," *New Mexico Historical Review* 83, no. 1 (2008): 21–22. Front-page story in the *Santa Fe New Mexican* by William McNughty.

5. Ibid.

6. "Impact Los Alamos Project," Rubén Montoya.

7. Charlotte Serber, "Labor Pains," in Serber and Wilson, eds., *Standing by and Making Do*, 57.

8. Betty Shouse and Marjorie Miller, "Open City," *New Mexico Magazine*, January 1958, 21–22.

9. Clarfield and Wiecek, *Nuclear America*, 204.

10. Fisher, *Los Alamos Experience*, 91.

11. Hunner, *Inventing Los Alamos*, 188.

12. Carl Abbott, "Building the Atomic Cities: Richland, Los Alamos, and the American Planning Language," in Hevly and Findlay, eds., *The Atomic West*, 106.

13. Daniel Lang, "A Reporter in New Mexico: Los Alamos," *The New Yorker*, April 1948, 68.

14. DeGroot, *The Bomb*, 281.

15. Niklaus and Feldman, *How Safe Is New Mexico's Atomic City?*, 52.

16. Lang, "A Reporter in New Mexico," 71.

17. Dimas Chavez, interview by Cindy Kelly, Washington, DC, February 13, 2013, Atomic Heritage Foundation and Los Alamos Historical Society, "Voices of the Manhattan Project."

18. Hunner, *Inventing Los Alamos*, 144.

19. Interview of Bill Jette in Mason, *Children of Los Alamos*, 60.

20. Joe Alex Morris, "The Cities of America: Los Alamos," *The Saturday Evening Post*, December 11, 1948, Ralph Carlisle Smith Papers on Los Alamos 1924–1957, Albuquerque: Center for Southwest Research, University Libraries, University of New Mexico, Collection MSS149BC, Box 1, Folders 28, 29, 30.

21. Gusterson, *Nuclear Rites*, 4.

22. Ibid., 88.

23. George Fitzpatrick, "Hush-hush no.2," *New Mexico Magazine*, March 1954, 47. Emphasis added.

24. On that day, Lt. Bob Carp was assigned the job of manually removing "the locking pin designed to prevent accidental in-flight release of bombs" during landing. The reason for the drop is disputed. It has been said that the plane bounced through a pocket of turbulent air, and Carp grabbed for a lever that triggered the drop. But Carp later asserted it was a "defectively designed manual release mechanism" that had been accidently pulled.

25. Les Adler, "Albuquerque's Near Doomsday," *The Albuquerque Tribune*, January 20, 1994, Center for Southwest Research Vertical Files, Albuquerque: University Libraries, University of New Mexico, Atomic Bomb—Dropped over Albuquerque.

26. Alexander, *History of Sandia Corporation*, 48.

27. Furman, *Sandia National Laboratories*, 224.

28. Sylvia Eberhart, "How the American People Feel About the Atomic Bomb," *Bulletin*, June 3, 1947, Ferenc M. Szasz Papers, 1894–2005, Albuquerque: Center for Southwest Research, University Libraries, University of New Mexico, Collection MSS552BC, Box 12, Folder 15, Articles on Nuclear Weapons, Fear of and Effects, 1946–1988.

29. Bartimus and McCartney, *Trinity's Children*, 81.

30. Scott McCartney, "Two Families, One Destiny: N.M. Sons Evicted by War Project now Are Brothers in Arms," *Albuquerque Tribune*, June 23, 1988, Center for Southwest Research Vertical Files, Albuquerque: University Libraries, University of New Mexico, Los Alamos, NM—Impact of Manhattan Project on Area Communities, Homesteaders on Pajarito Plateau, 1942. Also see, Bartimus and McCartney, *Trinity's Children*, 84–85.

31. Interview of Bill Jette in Mason, *Children of Los Alamos*, 62.

32. "Impact Los Alamos Project," Florida Martínez.

33. "Impact Los Alamos Project," Delfido Fernández, interview by Troy Fernández, Chimayó, NM, February 27, 1994, Box 1, CDs 36–37.

34. "Impact Los Alamos Project," Paul Emilio Frésquez, interview by Kenneth Salazar, La Mesilla, NM, April 24, 1995, Box 1, CDs 39–40.

35. "Salute to Los Alamos," *New Mexico Magazine*, April 1955, 37.

36. D. M. Brown, Major A.F., "Guided Missile Test Base Is Established at Alamogordo Field," *New Mexico Magazine*, January 1948, 50.

37. See Dickson, *Sputnik*, chap. 10, "Sputnik's Legacy."

38. Thomas J. Morris, Letter to Secretary William F. Darmitzel, Committee on Atomic Affairs, Office of the Governor, Santa Fe, NM, November 23,

1959, Governor John Burroughs Papers, 1959–1960, Santa Fe: New Mexico State Records Center and Archive, Collection 1959–239, Box 13191, Folder 24, Committee of Atomic Affairs (1959–1960). William F. Darmitzel, for the Committee on Atomic Affairs, Letter to Dennis Chavez, November 20, 1959, Governor John Burroughs Papers, 1959–1960, Santa Fe: New Mexico State Records Center and Archive, Collection 1959–239, Box 13191, Folder 24, Committee of Atomic Affairs (1959–1960).

39. Wood, "The Transformation of Albuquerque," 228.

40. Shroyer, *Secret Mesa*, 8–9.

41. Roman adage by Publius Flavius Vegetius Renatus (385–435 CE) in *Epitoma Rei Militaris* (Epitome of Military Science).

42. Shroyer, *Secret Mesa*, 17, 30, 33, 151.

43. Boyer, *By the Bomb's Early Light*, 114, 120, 296–300.

44. Jones, "A Review of the History of U.S. Radiation Protection Regulations, Recommendations, and Standards," 106; Inkret, Taschner, and Meinhold, "A Brief History of Radiation Protection Standards," 117.

45. Hales, *Atomic Spaces*, 278, 282, 289.

46. To read the document, see "The United States Strategic Bombing Survey: The Effects of the Atomic Bombings of Hiroshima and Nagasaki, June 30, 1946," Washington, DC: Government Printing Office, *Roger Williams University E-Books*, accessed October 2, 2014, http://docs.rwu.edu/cgi/viewcontent.cgi?article=1000&context=rwu_ebooks. Also see, "The Effects of the Atomic Bomb on Hiroshima, Japan," and "The Effects of the Atomic Bomb on Nagasaki, Japan," Physical Damage Division, May and June 1947, accessed November 10, 2014, https://archive.org/search.php?query=subject%3A%22U.S.%20Strategic%20Bombing%20Survey%22.

47. "Disclosure to Patients Injected with Plutonium," Summary Sheet SECY-75–130, August 13, 1974, 4, Ferenc M. Szasz Papers, 1894–2005, Albuquerque: Center for Southwest Research, University Libraries, University of New Mexico, Collection MSS552BC, Box 14, Folder 9, Radiation Poisoning, 1944–1993.

48. Edward Markey, Letter to Energy Secretary Hazel O'Leary, August 24, 1994, National Security Archive, Washington, DC: George Washington University, accessed March 22, 2018, https://nsarchive2.gwu.edu/radiation/dir/mstreet/commeet/meet6/brief6/tab_1/br6l1h.txt.

49. Welsome, *The Plutonium Files*.

50. The "atomic soldiers" were used as guinea pigs in the Nevada desert to test the psychological effects of nuclear warfare during the Desert Rock exercises between 1951 and 1957. Several incidents over the same decade brought fallout to the forefront of nuclear anxieties. In May 1953, the Dirty Harry test on the

Yucca Flat at NTS blanketed residents of Saint George, Utah, with as much radiation as nuclear workers were allowed in a year. One year later, the Castle Bravo test, the largest US explosion at Bikini, entailed the Lucky Dragon incident, which dealt a massive blow to the practice of atmospheric testing. See Adriana Rowberry, "Castle Bravo: The Largest U.S. Nuclear Explosion," Brookings, February 27, 2014, accessed October 4, 2014, http://www.brookings.edu/blogs/up-front/posts/2014/02/27-castle-bravo-largest-us-nuclear-explosion-rowberry; and Moss, ed., *Men Who Play God*, 87–88.

51. Turkevich, "Assuring Public Safety in Continental Weapons Tests," 88.

52. "Alice Stewart," *The Guardian*, June 28, 2002, accessed February 28, 2015, http://www.theguardian.com/news/2002/jun/28/guardianobituaries.nuclear. Also see, Greene, *The Woman Who Knew Too Much*.

53. Inkret, Taschner, and Meinhold, "A Brief History of Radiation Protection Standards," 120.

54. "Radiation Standards Causing Confusion," in Niklaus, and Feldman, *How Safe Is New Mexico's Atomic City?*, 15–17.

55. Clarfield and Wiecek, *Nuclear America*, 443.

56. McKee, *The Zia Company*, 34–35.

57. "Impact Los Alamos Project," Josefita Velarde.

58. "Impact Los Alamos Project" Genaro Martínez.

59. "Impact Los Alamos Project," Rubén Montoya.

60. The first lawsuit was filed by a Massachusetts woman, the widow of an employee who had died of a tumor of the lymphatic system. The second lawsuit was "brought by Saul Bramer, 57, an employee of TRW Nuclear Systems Inc., a California defense contractor" who was at the lab during the 1971 accidental plutonium release. Bramer filed a claim for damages first against the AEC and then against the ERDA, but the courts ruled against him.

61. "Atomic Mishap Ever Present Danger," in Niklaus and Feldman, *How Safe Is New Mexico's Atomic City?*, 53–55.

62. "Impact Los Alamos Project," Jasper Tucker, interview by Dot Waldrip, Dixon, NM, November 12, 1995, Box 2, CDs 38–39.

63. "Impact Los Alamos Project," José Benito Montoya.

64. "Impact Los Alamos Project," Pedro Martínez.

65. "Impact Los Alamos Project," Ramón Frésquez.

66. "Impact Los Alamos Project," Alfonso Mascarenes.

67. Juan Estevan Arellano, "Oral History Program Examines Impact of Los Alamos National Lab on Paisanos," *Land: Different Values*, Winter 1993–1994, Center for Southwest Research Vertical Files, Albuquerque: University Libraries, University of New Mexico, Los Alamos, NM—Impact of Manhattan Project on Area Communities, Homesteaders on Pajarito Plateau, 1942.

68. "Impact Los Alamos Project," Richard Cook. Emphasis added.

69. "Impact Los Alamos Project," Paul Emilio Frésquez.

70. "Impact Los Alamos Project," Larry Dillon, interview by Homer Campbell, April 23, 1995, Box 1, CD 35.

71. "Impact Los Alamos Project," Charles Montaño.

Chapter Nine

1. Kevin J. Fernlund, "Mining the Atom: The Cold War Comes to the Colorado Plateau, 1948–1958," *New Mexico Historical Review* 69, no. 4 (October 1994): 345–48.

2. Allan Richard, "Hold the Line, Reflection on a Nuclear Drama," *Taos Magazine*, Winter 1979–80, 16, Center for Southwest Research Vertical Files, Albuquerque: University Libraries, University of New Mexico, Technology—NM—Impact on Earth.

3. The most recent and thorough work on the subject is journalist Vincent B. Price's book *The Orphaned Land: New Mexico's Environment since the Manhattan Project.*

4. US National Research Council and Committee, "The Nuclear Weapons Complex," vii.

5. Ibid., 10–11.

6. National Research Council and Committee, "The Nuclear Weapons Complex," 43, 34–35.

7. Szasz, *Larger Than Life*, 48; Burt Hubbard, "N. M. Research Leaves Radioactive Residue," *Albuquerque Tribune*, February 3, 1981, A-2, Center for Southwest Research Vertical Files, Albuquerque: University Libraries, University of New Mexico, Technology—NM—Impact on Earth.

8. "Impact Los Alamos Project, " Ramón Frésquez.

9. "Impact Los Alamos Project," Paul Emilio Frésquez.

10. Niklaus and Feldman write, "One study reported in 1973 to a meeting of the International Radiation Protection Association in Washington, DC, noted that plutonium-238 and 239 concentrations in the lung and hide of rodents sampled in one Los Alamos canyon suggested that windblown soil particles may be a prime contamination mechanism. . . . Cesium-137 was also discovered to be elevated in mule deer inhabiting the canyon areas, with one deer exhibiting concentrations of the radionuclide in muscle about 35 times higher than deer in non-contaminated areas. . . . Elevated tritium concentrations two to five times normal were also observed in mule deer, ravens, and stellar jays collected from the canyon area." "LASL Scientists Keep Eye on Radiation's Spread," in Niklaus

and Feldman, *How Safe Is New Mexico's Atomic City?*, 28.

11. Niklaus and Feldman, *How Safe Is New Mexico's Atomic City?*, 1.

12. Editorial, "LASL's Waste Woes," in Niklaus and Feldman, *How Safe Is New Mexico's Atomic City?*, 62.

13. Burt Hubbard, "N.M. Research Leaves Radioactive Residue," *Albuquerque Tribune*, February 3, 1981, A-2, Center for Southwest Research Vertical Files, Albuquerque: University Libraries, University of New Mexico, Technology—NM—Impact on Earth.

14. "Radiation: How Much Is too Much?," in Niklaus and Feldman, *How Safe Is New Mexico's Atomic City?*, 9–13.

15. Scott McCartney, "A Legacy of Deadly Secrets," Insights, *The Albuquerque Tribune*, November 19, 1991, Center for Southwest Research Vertical Files, Albuquerque: University Libraries, University of New Mexico, Los Alamos, NM—History of, General.

16. Raam Wong, "Bomb Work Dumping Confirmed, UC Acknowledges Radioactive Waste Released During Manhattan Project," June 18, 2008, 1–2, Fray Angélico Chávez History Library Vertical Files, Santa Fe, NM: Palace of the Governors, Department of Cultural Affairs, Hazardous waste 1.

17. Ibid.

18. Mark Oswald, "Costs Force Plaintiff to Drop LANL Lawsuit," *Albuquerque Journal*, May 18, 2010, accessed December 19, 2014, http://www.abqjournal.com/news/state/18234661559newsstate05-18-10.htm.

19. Hubbard, "N.M. Research Leaves Radioactive Residue," A-2; Stacey Matlock, "A History of Innovation and Dysfunction," *Santa Fe New Mexican*, January 2, 2016, available at http://www.santafenewmexican.com/news/local_news/a-history-of-innovation-and-dysfunction-at-los-alamos-national/article_6bde4aee-077f-56a6-836b-eab1d289271e.html.

20. Ben Neary, "EPA Postpones Cleanup Hearing at Los Alamos," July 20, 1989, 3, Fray Angélico Chávez History Library Vertical Files, Santa Fe, NM: Palace of the Governors, Department of Cultural Affairs, Hazardous waste 1.

21. Masco, *The Nuclear Borderlands*, 27, 32.

22. Hubbard, "N.M. Research Leaves Radioactive Residue," A-2.

23. "Cancer Rate Elevated in Los Alamos County," in Niklaus and Feldman, *How Safe Is New Mexico's Atomic City?*, 45, 42.

24. Ibid., 48.

25. According to the UNM Cancer Center, Los Alamos County ranked first in New Mexico for childhood cancer death rates between 1953 and 1987, and according to the New Mexico Department of Health, the brain cancer rate from 1984 to 1988 was twice the national rate. Price, *The Orphaned Land*, 190–91.

26. Interview of Betty Marchi Schulte in Mason, *Children of Los Alamos*, 74.

27. New Mexico Museum of Art, "History: The Politics of Water," New Mexico Art Tells New Mexico History, *New Mexico Museum of Art*, Santa Fe: New Mexico Museum of Art, 2010, accessed March 11, 2015, http://online.nmart museum.org/nmhistory/people-places-and-politics/water/history-water .html. Also see Ira G. Clark, *Water in New Mexico*.

28. "Report: LANL Contaminating Nearby Areas," *Santa Fe New Mexican*, June 8, 1988, A-2, Fray Angélico Chávez History Library Vertical Files, Santa Fe, NM: Palace of the Governors, Department of Cultural Affairs, Hazardous waste 1.

29. John Holmes, "Nuclear Waste: You Can't Toss It in the Garbage," *Albuquerque Journal Magazine*, IMPACT, July 17, 1979, 8, Fray Angélico Chávez History Library Vertical Files, Santa Fe, NM: Palace of the Governors, Department of Cultural Affairs, Hazardous waste 1.

30. Price, *Orphaned Land*, 116.

31. "State-Designated Top-Priority National Priorities List (NPL) Sites," US Environmental Protection Agency, accessed March 22, 2018, https://www.epa. gov/superfund/state-designated-top-priority-national-priorities-list-npl-sites. For detail on the progress of remedial operations at the site, see "South Valley Albuquerque, NM: Cleanup Activities," US Environmental Protection Agency, accessed March 22, 2018, https://cumulis.epa.gov/supercpad/SiteProfiles/ index.cfm?fuseaction=second.cleanup&id=0600881.

32. Price, *The Orphaned Land*, 24–25.

33. Burt Hubbard, "Landfill 'Time Bombs' Ticking Throughout New Mexico," *Albuquerque Tribune*, January 30, 1981, A-10, Center for Southwest Research Vertical Files, Albuquerque: University Libraries, University of New Mexico, Technology—NM—Impact on Earth.

34. Price, *The Orphaned Land*, 105–6, 170, 97.

35. Hubbard, "N.M. Research Leaves Radioactive Residue."

36. Amundson, *Yellowcake Towns*, 162. There were also 24 million tons of waste at the Anaconda plant at Bluewater; 1.24 million tons at the United Nuclear-Homestake-New Mexico Partners tailings pile; 20 million tons at the Homestake-Spin mill; and 2.6 million tons at the Phillips mill.

37. LaDuke Westigaard, "Uranium Mines on Native Land: The New Indian Wars," *Harvard Crimson*, May 2, 1969, accessed August 26, 2015, http://www .thecrimson.com/article/1979/5/2/uranium-mines-on-native-land-pthe/.

38. "Hazardous Waste—N.M.'s Lethal Legacy," *Albuquerque Tribune*, Albuquerque, January 26, 1981, Center for Southwest Research Vertical Files, Albuquerque: University Libraries, University of New Mexico, Technology—NM—Impact on Earth.

39. Szasz, *Larger Than Life*, 48.

40. Linda M. Richards, "On Poisoned Ground," *Distillations*, Science History Institute, Spring 2013, accessed March 22, 2018, https://www.sciencehistory.org/distillations/magazine/on-poisoned-ground.

41. Marley Shebala, "Poison in the earth," *Navajo Times*, July 23, 2009, accessed March 13, 2015, http://navajotimes.com/news/2009/0709/072309 uranium.php#.VQLmTOEf0ao.

42. Richards, "On Poisoned Ground."

43. See Visgilio and Whitelaw, eds., *Our Backyard*.

44. "State & County QuickFacts New Mexico," US Census Bureau, Washington, DC: US Department of Commerce, accessed October 23, 2014, http://quickfacts.census.gov/qfd/states/35000.html.

45. See De la Garza, Kruszewski, and Arciniega, *Chicanos and Native Americans*.

46. Price, *The Orphaned Land*, 12, 33, 19, 116, 96.

47. "Impact Los Alamos Project," Paul Frésquez, interview by Kenneth Salazar, March 6, 1995, Box 1, CDs 38–41.

48. "Impact Los Alamos Project," Ramón Frésquez.

49. Interview in Masco, *The Nuclear Borderlands*, 192–93.

50. "Impact Los Alamos Project," C. L. Hunter.

51. "Impact Los Alamos Project," Leroy Martínez, interview by Carlos Vásquez, Chimayó, NM, November 3, 1991, Box 2, CD 64.

52. "Impact Los Alamos Project," Symposia for the Community, UNM Division of Continuing Education, 1996, Box 2, CD 51.

53. Dick Layman, "Latino Poll Shows Support for Conservation Efforts," *Public News Service*, Boulder, CO: Public News Service, September 21, 2011, accessed November 4, 2014, http://www.publicnewsservice.org/2011-09-21/public-lands-wilderness/latino-poll-shows-support-for-conservation-efforts/a22285-1.

54. Interview in Masco, *The Nuclear Borderlands*, 194.

55. Kuletz, *The Tainted Desert*, 7.

56. Ibid., xviii, 8, 43.

57. Churchill and LaDuke, "Native America."

58. Nukewatch, Santa Fe, NM: Nuclear Watch New Mexico, accessed March 22, 2018, https://www.nukewatch.org/index.html#.WrPiKajwaUk. See also Kay Matthews's online magazine of environmental politics in New Mexico, *La Jicarita*, https://lajicarita.wordpress.com.

59. William D. Purtymun and William R. Kennedy, "Geology and Hydrology of Mesita del Buey," LA-4660 UC-41 Health and Safety TIF-4500 (Los Alamos, NM: Los Alamos Scientific Laboratory of the University of California, November

1970–May 1971), accessed April 19, 2015, http://www.osti.gov/scitech /servlets/purl/4044830/. "Material Disposal Areas," *Los Alamos National Laboratory*, Los Alamos National Security, Department of Energy's National Nuclear Security Administration, accessed March 22, 2018, https://www.lanl. gov/environment/cleanup/sites-projects/material-disposal-areas.php.

60. Advisory Committee Staff, "Fact Sheet on Radiolanthanum (Rala) Tests," Memo to the Members of the Santa Fe Small Panel, National Security Archive (Washington, DC: George Washington University, January 10, 1995), accessed March 22, 2018, https://nsarchive2.gwu.edu/radiation/dir/mstreet/commeet /pm03/pm3brf/tab_b/pm03b1.txt.

61. Masco, *The Nuclear Borderlands*, 135–38.

62. Advisory Committee on Human Radiation Experiments, *The Human Radiation Experiments: Final Report*, 332–33.

63. "LASL Aide Admits Waste Report 'Misleading,'" in Niklaus and Feldman, *How Safe Is New Mexico's Atomic City?*, 41.

64. Price, *The Orphaned Land*, 193.

65. Darryl Martínez, interview by the author, Governor's Office of San Ildefonso Pueblo, NM, September 17, 2013.

66. Masco, *The Nuclear Borderlands*, 141.

67. Ross, "One Mother Earth, One Doctor Water," 110, 138, 139. Such a statement calls to mind psychiatrist and philosopher Frantz Fanon's analysis of the cultural assimilation of colonized people. He writes, "The first thing the colonial subject learns is to remain in his place and not overstep its limits." Fanon, *Les Damnés de la terre*, 58.

68. "LASL Aide Admits Waste Report 'Misleading,'" in Niklaus and Feldman, *How Safe Is New Mexico's Atomic City?*, 41.

69. Naomi Archuleta, interview by the author, Office of Environmental Affairs of Ohkay Owingeh Pueblo, NM, September 17, 2013.

70. Ross, "One Mother Earth, One Doctor Water," 168.

71. Eichstaedt, *If You Poison Us*, 164.

72. See Judy Pasternak, *Yellow Dirt*.

73. Ibid.

74. Eichstaedt, *If You Poison Us*, 43.

75. Westigaard, "Uranium Mines on Native Land."

76. Brugge, Benally, and Yazzie-Lewis, eds., *The Navajo People and Uranium Mining*, 12, 80, 50, 101, 122, 131.

77. Westigaard, "Uranium Mines on Native Land."

78. The six branches were the Departments of Energy Interior, and Agriculture; the Nuclear Regulatory Commission; Tennessee Valley Authority; and the Environmental Protection Agency.

79. Westigaard, "Uranium Mines on Native Land."

Chapter Ten

1. Charles W. Mills, *The Power Elite*, 7.

2. Hales, *Atomic Spaces*, 2, 5.

3. McDonald, *The New Mexican Economy*, 22.

4. Welsh, "The Land of Extremes: The Economy of Modern New Mexico," in Etulain, ed., *Contemporary New Mexico, 1940–1990*, 70.

5. Adelamar et al., "Poverty in New Mexico," 3–5.

6. This is observable in a series of reports issued by New Mexico State University that studied the economic impacts of DOE activities in the state. The reports study activities of the laboratories but also those of a biomedical and environmental research institute, a national waste repository, a national remedial action project, and several energy research and conservation programs.

7. Cohen and New Mexico State University, "The Economic Impact of Los Alamos National Laboratory and Sandia National Laboratories on the State of New Mexico Fiscal Year 1990," 6, 12.

8. US Census Bureau, "Poverty: 2008 and 2009," *American Community Survey Briefs*, issued September 2010, accessed March 22, 2018, https://www.census.gov/prod/2010pubs/acsbr09-1.pdf.

9. Adelamar et al., "Poverty in New Mexico," 56, 89–98.

10. Dietz, "The Impact of Los Alamos National Laboratory on Northern New Mexico," 105–6.

11. Adelamar et al., "Poverty in New Mexico," 56, 89–98.

12. "Impact Los Alamos Project," Hipólita Fernández.

13. "Impact Los Alamos Project," Danny Martínez.

14. "Impact Los Alamos Project," Patricia Trujillo-Oviedo.

15. Don E. Alberts, "Kirtland Air Force Base: Its Origin and Activities," Research Reports, History of Albuquerque Exhibits Series, Vol. 5, The Albuquerque Museum, History Division, March 1981, in Karafantis, "Weapons Labs and City Growth," 128.

16. "Impact Los Alamos Project," Danny Martínez.

17. Walter Briggs, "Los Alamos, Mysteries, Past and Future," *New Mexico Magazine*, February 1969, 23.

18. Juan Estevan Arellano, "La Querencia: La Raza Bioregionalism," *New Mexico Historical Review* 72, no. 1 (January 1997): 32.

19. Dietz, "The Impact of Los Alamos National Laboratory on Northern New Mexico," 102.

20. Hevly and Findlay, *The Atomic West*, 105.

21. Tom Van Riper, "America's Richest Counties," Forbes.com, April 25, 2013, accessed January 31, 2014, http://www.forbes.com/sites/tomvanriper/2013/04/25/americas-richest-counties/. Also see Bishaw, *Poverty*, 3.

22. See "Poverty Rates of School-Age Population by County, 2013," US Census Bureau Press Release CB14–229, "Census Bureau Estimates Show How School-Age Child Poverty in Every County Compares with Prerecession Levels," Washington, DC: US Department of Commerce, December 17, 2014, accessed March 16, 2015, http://www.census.gov/newsroom/press-releases/2014/cb14-229.html.

23. Shroyer, *Secret Mesa*, 2–3.

24. US Economic Development Administration, "Public Works and Economic Development Act of 1965, as Amended ("PWEDA")," accessed March 22, 2018, https://www.eda.gov/pdf/about/Comprehensive-PWEDA.pdf.

25. North Central New Mexico Economic Development District, "Regional Development Plan for the North Central New Mexico Economic Development District," Santa Fe, NM, June 1977.

26. "Impact Los Alamos Project," Harold Gibson.

27. Mason, *Children of Los Alamos*, xii.

28. Interview of Secundino Sandoval in Mason, *Children of Los Alamos*, 17.

29. See Manuel García y Griego, "The Importation of Mexican Contract Laborer to the United States, 1942–1964," in Gutiérrez, ed., *Between Two Worlds*, 45–85.

30. Mason Sutherland and Justin Locke, "Adobe New Mexico," *National Geographic Magazine*, Tampa, FL, December 1949, 825, Ralph Carlisle Smith Papers on Los Alamos 1924–1957, Albuquerque: Center for Southwest Research, University Libraries, University of New Mexico, Collection MSS-149BC, Box 1, Folder 62.

31. Interview of Nella Fermi Weiner in Mason, *Children of Los Alamos*, 78.

32. "Impact Los Alamos Project," Senni A. Gallegos.

33. "Impact Los Alamos Project," Santiago Bustamente, interview by Troy Fernández, Santa Fe, NM, August 8, 1995, Box 1, CD 28.

34. Dietz, "The Impact of Los Alamos National Laboratory on Northern New Mexico," 90–91.

35. "Impact Los Alamos Project," Paul Emilio Frésquez.

36. Chuck Montaño, "60 Years and the Rent's Still Owed," *Advocacy*, Center for Southwest Research Vertical Files, Albuquerque: University Libraries, University of New Mexico, Los Alamos, NM—Impact of Manhattan Project on Area Communities, Homesteaders on Pajarito Plateau, 1942.

37. "Impact Los Alamos Project," Maxine Beckman.

38. Mary A. Meyer, "Jobs, Work-Related Values, and Attitudes of Staff and Technicians at a National Laboratory" (PhD diss., University of New Mexico, Department of Anthropology, 1984), 16.

39. "Impact Los Alamos Project," Senni A. Gallegos.

40. Dietz, "The Impact of Los Alamos National Laboratory on Northern New Mexico," 102.

41. Córdova, *The 3 1/2 Cultures of Española*, 18, 49–52.

42. Dietz, "The Impact of Los Alamos National Laboratory on Northern New Mexico," 102.

43. Córdova, *The 3 1/2 Cultures of Española*, 49–52.

44. "Impact Los Alamos Project," C. L. Hunter.

45. "Impact Los Alamos Project," Starr Beckman, interview by Linda Campbell, April 18, 1995, Box 1, CD 27.

46. "Impact Los Alamos Project," Charles Montaño. Six months after Montaño arrived at the lab in 1979, his friend, Mike Grallegos, with whom he shared an office, committed suicide at age twenty-seven.

47. Nikolas Athanassios Stergioulas, "An Analysis of the Structural Causes of Poverty in New Mexico" (Master's thesis, University of New Mexico, 1979), v–vi.

48. Adelamar et al., "Poverty in New Mexico," 36.

49. Barrera, *Race and Class in the Southwest*, 38, 43.

50. Hales, *Atomic Spaces*, 193.

51. Masco, *The Nuclear Borderlands*, 132–33.

52. "Impact Los Alamos," Vásquez, 7–8.

53. Rothman, *On Rims and Ridges*, 277.

54. "Impact Los Alamos Project," Loyda Martínez. Emphasis added.

55. "Impact Los Alamos Project," Paul Frésquez.

56. Ross, "One Mother Earth, One Doctor Water," 126.

57. Dietz, "The Impact of Los Alamos National Laboratory on Northern New Mexico," 88–94.

58. "Impact Los Alamos Project," Rubén Montoya.

59. Welsh, "The Land of Extremes: The Economy of Modern New Mexico," in Etulain, ed., *Contemporary New Mexico, 1940–1990*, 74.

60. See Tijerina, *They Called Me "King Tiger."*

61. See Gardner, *Grito! Reies Tijerina and the New Mexico Land Grant War of 1967*.

62. Robert J. Rosenbaum and Robert W. Larson, "Mexicano Resistance to the Expropriation of Grant Lands in New Mexico," in Briggs and Van Ness, eds., *Land, Water, and Culture*, 270.

63. Rothman, *On Rims and Ridges*, 276.

64. Translation: He that loses his land loses his memory.

65. See Anil Rupasingha and J. Michael Patrick, "Rural New Mexico Economic

Conditions and Trends," CR-651, Las Cruces: New Mexico State University, College of Agricultural, Consumer and Environmental Sciences, Cooperative Extension Service, April 2012, accessed April 19, 2015, http://aces.nmsu.edu/ pubs/_circulars/CR-651.pdf, 2.

66. Gray, "Reclaiming Querencia," 479.

67. Arellano, "La Querencia," 32, 35.

68. Adam Rankin, "Heirs Say Lab Must Pay for Its Land Grab, Attorney Says Homesteaders Are United Behind Lawsuit," *Albuquerque Journal North*, May 5, 2003, 5, Fray Angélico Chávez History Library Vertical Files, Santa Fe, NM: Palace of the Governors, Department of Cultural Affairs, Los Alamos National Laboratories—Pajarito Plateau Hom 1.

69. "Impact Los Alamos Project," Leroy Martínez.

70. "Impact Los Alamos Project," Lucille Sanchez.

71. Dietz, "The Impact of Los Alamos National Laboratory on Northern New Mexico," 88–94.

72. Córdova, *The 3 1/2 Cultures of Española*, 53, 59.

73. Anne Marie Smith, *New Mexico Indians: Economic, Educational, and Social Problems* (Santa Fe: Museum of New Mexico Press, 1969), 22–23.

74. Both case studies are from Dietz, "The Impact of Los Alamos National Laboratory on Northern New Mexico," 88–94.

75. Stergioulas, "An Analysis of the Structural Causes of Poverty in New Mexico," 35.

76. Smith, *New Mexico Indians*, 89, 54.

77. Lange, *Cochití*, 190.

78. Fisher, *Los Alamos Experience*, 89.

79. Church, *The House at Otowi Bridge*, 2.

80. Kaa Fedeh interview in Ross, "One Mother Earth, One Doctor Water," 60, 75.

81. Interviews in Masco, *The Nuclear Borderlands*, 106, 110–11.

Chapter 11

1. Wills, *Bomb Power*, 8–9.

2. Welsome, *The Plutonium Files*, 482.

3. The Military Construction and Reserve Forces Facilities Authorization Acts and the Military Construction Act in 1973, and the Military Construction Appropriation Act in 1980.

4. Irving Porter, "Farm and Ranch Folks Project." Marc Simons, "The Last Stand of John Prather," *Prime Time*, April 2007, 4, Center for Southwest Research Vertical Files, Albuquerque: University Libraries, University of New Mexico, White Sands, NM—Monument, Missile Range, Ranchers' Protest.

5. "Skeen, Ranchers Testify on Compensation Bill," *News from Congressman Joe Skeen*, Press Release, Washington, DC, June 13, 1990, Fray Angélico Chávez History Library Vertical Files, Santa Fe, NM: Palace of the Governors, Department of Cultural Affairs, White Sands Missile Range 1.

6. "Farm and Ranch Folks Project," Ernest Aguayo.

7. Richard Pipes, "White Sands Reaches Out to Ranch Lands," Center for Southwest Research Vertical Files, Albuquerque: University Libraries, University of New Mexico, White Sands, NM—Monument, Missile Range, Ranchers' Protest.

8. Bob Groves, "The Atomic House that Schmidt Built," Impact, *Albuquerque Journal Magazine*, July 14, 1987, Center for Southwest Research Vertical Files, Albuquerque: University Libraries, University of New Mexico White Sands, NM—Monument, Missile Range, Ranchers' Protest.

9. New Mexico State University, "Values Figured for Ranches on White Sands Missile Range," February 8, 1983, 28, 29, Fray Angélico Chávez History Library Vertical Files, Santa Fe, NM: Palace of the Governors, Department of Cultural Affairs, White Sands Missile Range 1.

10. Bill Diven, "Angry Ranchers Take Up Vigil at White Sands," *Albuquerque Journal*, July 17, 1985, 2, Center for Southwest Research Vertical Files, Albuquerque: University Libraries, University of New Mexico, White Sands, NM—Monument, Missile Range, Ranchers' Protest.

11. "Farm and Ranch Folks Project," David McDonald.

12. Isabel Foreman, "Ranch House Part of U.S. Atomic History," *Albuquerque Journal*, September 9, 1984, D-1, Center for Southwest Research Vertical Files, Albuquerque: University Libraries, University of New Mexico, White Sands, NM—Monument, Missile Range, Ranchers' Protest.

13. Edgington, *Range Wars*, 170.

14. Ibid., 2.

15. "Farm and Ranch Folks Project," Ernest Aguayo.

16. Tina Córdova, "Statement of January 28, 2010," *Tularosa Basin Downwinders Consortium*, Santa Fe, NM: Southwest Research and Information Center, accessed April 9, 2015, http://www.sric.org/voices/2010/v11n2/TBDC .pdf.

17. Tom Udall, "Justice for Tularosa Basin Downwinders," Tom Udall Senator for New Mexico, Washington, DC, July 31, 2015, accessed September 10, 2015, https://www.tomudall.senate.gov/news/press-releases/justice-for-tularosa-basin-downwinders.

18. "Tijerina's Citizen's Arrest Attempt Unsuccessful," *Santa Fe New Mexican*, June 8, 1969.

19. William J. Clinton, "Statement of Administration Policy: H.R. 2538—Guadalupe-Hidalgo Treaty Land Claims Act (10 September 1998)," The

American Presidency Project, Gerhard Peters and John T. Woolley, accessed December 10, 2014, http://www.presidency.ucsb.edu/ws/?pid=74392.

20. Masco, *The Nuclear Borderlands*, 147.

21. Kathleene Parker, "Hispanic Claims Ignored in Los Alamos Deal," Opinion, *The Santa Fe New Mexican*, January 11, 2003, A-5, Fray Angélico Chávez History Library Vertical Files, Santa Fe, NM: Palace of the Governors, Department of Cultural Affairs, Los Alamos National Laboratories—Pajarito Plateau Hom 1. Jim Yardley, "Hispanic Heirs Seek Reparations for N.M. Lab Land," *Albuquerque Tribune*, January 2000, A-5, Center for Southwest Research Vertical Files, Albuquerque: University Libraries, University of New Mexico, Los Alamos, NM—Impact of Manhattan Project on Area Communities, Homesteaders on Pajarito Plateau, 1942.

22. Chuck Moutain, "Los Alamos Fire May Bring Heat and Justice to Original Latino Families," *Imagen Magazine*, September 2000, 25, Center for Southwest Research Vertical Files, University Libraries, University of New Mexico, Los Alamos, NM—Health Hazards.

23. Diana Heil, "Justice Draws Near for Heirs of Land Taken by U.S. Government," *Santa Fe New Mexican*, October 12, 2004, B1, Fray Angélico Chávez History Library Vertical Files, Santa Fe, NM: Palace of the Governors, Department of Cultural Affairs, Los Alamos National Laboratories—Pajarito Plateau Hom 1.

24. Congressional Budget Office Cost Estimate, "S.1773 Pueblo de San Ildefonso Claims Settlement Act of 2005," April 19, 2006, accessed December 11, 2014, http://www.cbo.gov/sites/default/files/s1773.pdf. "House OKs San Ildefonso Land Claim Settlement," *Albuquerque Journal*, September 13, 2006, accessed December 11, 2014, http://www.abqjournal.com/news/state/apclaim09-13-06.htm.

25. Córdova, *Children of the Pajarito Plateau*, 265.

26. Juan Esteban Arellano, "Oral History Program Examines Impact of Los Alamos National Lab on *Paisanos*," Winter 1993–1994, Center for Southwest Research Vertical Files, Albuquerque: University Libraries, University of New Mexico, Los Alamos, NM—Impact of Manhattan Project on Area Communities, Homesteaders on Pajarito Plateau, 1942.

27. Wren Propp, "Suit: Nuke Project Harmed Hispanics, Mistreatment in 1940s Alleged," *Albuquerque Journal North*, February 10, 2001, 1–2, Fray Angélico Chávez History Library Vertical Files, Santa Fe, NM: Palace of the Governors, Department of Cultural Affairs, Los Alamos National Laboratories—Pajarito Plateau Hom 1.

28. Betty Childers, Judicial Affairs Writer, "Los Alamos Losing Ground in Sex-Bias Suit," *Albuquerque Journal*, April 29, 1984, B-11, Fray Angélico

Chávez History Library Vertical Files, Santa Fe, NM: Palace of the Governors, Department of Cultural Affairs, Los Alamos National Laboratories—Employee Relations 1.

29. National Nuclear Security Administration, "Maintaining the Stockpile," accessed March 25, 2015, http://nnsa.energy.gov/aboutus/ourprograms/ defenseprograms/aboutdefenseprograms.

30. Los Alamos National Laboratory, "Budget," Los Alamos: Los Alamos National Security, accessed February 18, 2015, http://www.lanl.gov/about/ facts-figures/budget.php.

31. "Los Alamos Laboratory Reaches $625,000 Settlement to Remedy Discrimination Against Hispanic Workers," OPA Press Release, Washington, DC: US Department of Labor, May 13, 1988, accessed December 17, 2014, http:// www.dol.gov/opa/media/press/opa/archive/opa98210.htm.

32. "LANL Accused of Racial Discrimination," August 28, 2004, B-1, Fray Angélico Chávez History Library Vertical Files, Santa Fe, NM: Palace of the Governors, Department of Cultural Affairs, Los Alamos National Laboratories—Layoffs 1.

33. "If You Are a Woman or Hispanic and Were Employed at the Los Alamos National Laboratory on or after December 10, 2000, a Proposed Class Action Settlement May Affect Your Rights," *Albuquerque Journal*, August 11, 2006, A-10, Fray Angélico Chávez History Library Vertical Files, Santa Fe, NM: Palace of the Governors, Department of Cultural Affairs, Los Alamos National Laboratories—Layoffs 1.

34. Adam Rankin, "Heirs Say Lab Must Pay for Its Land Grab, Attorney Says Homesteaders Are United Behind Lawsuit," *Albuquerque Journal North*, May 5, 2003, 5, Fray Angélico Chávez History Library Vertical Files, Santa Fe, NM: Palace of the Governors, Department of Cultural Affairs, Los Alamos National Laboratories—Pajarito Plateau Hom 1.

35. John Arnold, "LANL Lawsuits to Cost UC $12M," *Albuquerque Journal*, May 20, 2006, accessed March 22, 2018, https://www.abqjournal.com /lanl/461588nm05-20-06.htm and Sarah Welsh, "Lab Workers Awarded $12 Million," *Rio Grande Sun*, December 17, 2009, http://www.riograndesun. com/northern_briefs/lab-workers-awarded-million/article_24afbafb-69b2 -5918-80ae-b3b1967fd179.html.

36. "Whistleblower Tells His Side of the Story; LANL: Former Auditor Speaks out a Year after Settling Lawsuit," Los Alamos Monitor Online, January 18, 2012, accessed December 18, 2014, http://www.lamonitor.com/ content/whistleblower-tells-his-side-story. Also see Amy Goodman and Juan González, "Birthplace of Atomic Bomb, New Mexico Remains Center of Massive U.S. Nuclear Arsenal," *Democracy Now!*, October 11, 2012,

accessed January 02, 2015, http://www.democracynow.org/2012/10/11/birthplace_of_atomic_bomb_new_mexico.

37. "Impact Los Alamos Project," Charles Montaño.

38. Mike Scarborough, email to the author, February 24, 2016. Scarborough is also the author of *Trespassers on Our Own Land*.

39. Staci Matlock, "LANL Faces Penalties in Cleanup Delays," *Santa Fe New Mexican*, June 27, 2014, accessed December 19, 2014, http://www.santafenewmexican.com/special_reports/from_lanl_to_leak/lanl-faces-penalties-in-cleanup-delays/article_fa31eb3c-fd36-5209-97d6-e531dbae436f.html.

40. Jim Ludwick and John Fleck, "Chavez to Texas: Hands Off Sandia," *Albuquerque Journal*, March 16, 2002, A-1, Fray Angélico Chávez History Library Vertical Files, Santa Fe, NM: Palace of the Governors, Department of Cultural Affairs, Sandia National Laboratories 1.

41. "Hispano Chamber Honors Sandia, Lockheed and TVC," *Albuquerque Journal*, Business Outlook, February 10, 2003, 8, Fray Angélico Chávez History Library Vertical Files, Santa Fe, NM: Palace of the Governors, Department of Cultural Affairs, Sandia National Laboratories 1.

42. Michael Hartranft, "Sandia Primes N.M.'s Economic Pump," *Albuquerque Journal*, Business Outlook, February 14, 2011, 4, Fray Angélico Chávez History Library Vertical Files, Santa Fe, NM: Palace of the Governors, Department of Cultural Affairs, Sandia National Laboratories 1.

43. Tim Anderson, "Mesa del Hole," *Alibi*, 28 November—4 December 2002, 28, Center for Southwest Research Vertical Files, Albuquerque: University Libraries, University of New Mexico, Technology—NM—Impact on Earth.

44. "Long-Term Stewardship Resource Center," Energy.gov, Washington, DC: US Department of Energy Office of Environmental Management, accessed January 2, 2015, http://energy.gov/em/services/communication-engagement/long-term-stewardship-resource-center.

45. "CANM Final Report: MWL Disposal of High Level Nuclear Waste," *Citizen Action New Mexico*, January 5, 2015, accessed March 22, 2018, http://www.radfreenm.org/index.php/mm-mwl/mwl-regulatory/mwl-nmed/citizen-action-vs-nmed/canm-final-mwl-report.

46. Stephanie Hiller, "Cover-up and Collusion at the Sandia National Laboratory Corral," *La Jicarita*, April 11, 2013, accessed December 19, 2014, https://lajicarita.wordpress.com/2013/04/11/cover-up-and-collusion-at-the-sandia-national-laboratory-corral/.

47. Udall, "Justice for Tularosa Basin Downwinders."

48. "Superfund Site Progress Profile Homestake Mining Co.," US Environmental Protection Agency, Washington, DC, accessed March 25, 2015, http://cumulis.epa.gov/supercpad/cursites/csitinfo.cfm?id=0600816.

49. "United Nuclear Corp. Mill Site, New Mexico," US Environmental Protection Agency, Washington, DC, accessed March 25, 2015, http://www.epa.gov/Region6/6sf/newmexico/united_nuclear/index.html.

50. Eichstaedt, *If You Poison Us,* 129, xvi.

51. "National Priorities List, NPL Site Narrative for Jackpile-Paguate Uranium Mine," and "Superfund Site Progress Profile Jackpile-Paguate Uranium Mine," US Environmental Protection Agency, Washington, DC, accessed October 24, 2014, http://www.epa.gov/superfund/sites/npl/nar1865.htm, and http://cumulis.epa.gov/supercpad/cursites/csitinfo.cfm?id=0607033.

52. See *Dii'go to Baahane, Four Stories About Water,* Dir. Deborah Begel and David Lindblom, sponsored by Eastern Navajo Diné Against Uranium Mining, Sierra Club Environmental Justice Office, Southwest Research and Information Center, UNM Community Environmental Health Program, and Connecting Higher Education Indigenously, Prod. Deborah Begel, 2012.

53. The five companies are Energy Fuels & Strathmore Minerals, Uranium Resources, Uranium International, Cotter, and Laramide Resources. Strathmore Minerals, which was taken over by Energy Fuels in 2013, had submitted a mining permit application in October 2009 for Roca Honda in the Grants mineral belt. The company also had exploration projects in Marquez, Church Rock, and Nose Rock. Uranium Resources (URI) sought to buy Rio Algom Mining in 2007 with uranium properties and a licensed mill site at Ambrosia Lake where it planned to build a new mill, but the deal was aborted in 2008. URI subsidiary Hydro Resources was licensed in 1994 to mine the Crownpoint and Church Rock ISL sites in New Mexico. After years of opposition, the license was validated by the Nuclear Regulatory Commission in 2006 and reactivated in 2011. Cotter, a General Atomic subsidiary, expected to treat ore from Mount Taylor at the rebuilt Canon City mill by 2014. Last, Laramide Resources had the La Jara Mesa project in the Grants mineral belt. "U.S. Uranium Mining and Exploration," World Nuclear Association, accessed December 25, 2014, http://www.world-nuclear.org/info/Country-Profiles/Countries-T-Z/Appendices/US-Nuclear-Fuel-Cycle-Appendix-1—US-Uranium-Mining-and-Exploration-/.

54. "In Situ Leach (ISL) Mining of Uranium," World Nuclear Association, accessed March 25, 2015, http://www.world-nuclear.org/info/Nuclear-Fuel-Cycle/Mining-of-Uranium/In-Situ-Leach-Mining-of-Uranium/.

55. Leona Morgan, "ENDAUM Statement to the New Mexico Indian Affairs Committee," Gallup, NM, Eastern Navajo Diné Against Uranium Mining, October 27, 2011, accessed January 2, 2015, http://www.nmlegis.gov/lcs/handouts/IAC%20110111%20Statement%20Leona%20Morgan%20ENDAUM.pdf; Amy Goodman and Juan González, "After Decades of Uranium Mining, Navajo Nation Struggles With Devastating Legacy of Contamination," Democracy

Now!, October 11, 2012, accessed January 2, 2015, http://www.democracynow
.org/2012/10/11/after_decades_of_uranium_mining_navajo.

56. These facilities include LANL, the Rocky Flats Plant, Idaho National
Engineering Laboratory, the Hanford Site, the Lawrence Livermore National
Laboratory, the Nevada Test Site, Argonne National Laboratory, Mound Labo-
ratories, Oakridge National Laboratory, and the Savannah River Site.

57. TRU are waste containing over 100 nanocuries of alpha-emitting transuranic isotopes per gram with half-lives greater than twenty years. Most of these
are plutonium-contaminated debris, rags, protective clothing, laboratory equipment, tools, soils, residues, and other materials used in the research, development, and fabrication testing of nuclear weapons. Only waste produced since
1970 can be shipped to WIPP.

58. US Department of Energy, "Waste Isolation Pilot Plant," 5, 14, Center
for Southwest Research Vertical Files, Albuquerque: University Libraries, University of New Mexico, Energy—NM—Nuclear—WIPP, 16.

59. Krens and Sankey, "WIPP and the Local Communities."

60. McCutcheon, *Nuclear Reactions*, 12.

61. Douglas W. Clark, "The Hot Topic of WIPP," *Quantum* 7, no. 1 (Spring
1990), Center for Southwest Research Vertical Files, Albuquerque: University
Libraries, University of New Mexico, Energy—NM—Nuclear—WIPP.

62. "Northern Pueblos Council Adopts Anti-WIPP Resolution," *Albuquer-
que Journal*, March 30, 1982, Center for Southwest Research Vertical Files,
Albuquerque: University Libraries, University of New Mexico, Energy—NM—
Nuclear—WIPP.

63. "Hands United to Stop WIPP," Flyer, Ferenc M. Szasz Papers, 1894–
2005, Albuquerque: Center for Southwest Research, University Libraries, University of New Mexico, Collection MSS552BC, Box 4, Folder 80, Waste Isolation Pilot Plant (WIPP), 1991.

64. "WIPP," New Mexico Department of Homeland Security and Emergency Management, Santa Fe, NM, accessed March 25, 2015, http://www
.nmdhsem.org/wipp.aspx.

65. "WIPP Route Goes Through Heart of City," *Albuquerque Journal*,
March 1, 1998, A-10, Center for Southwest Research Vertical Files, Albu-
querque: University Libraries, University of New Mexico, Energy—NM—
Nuclear—WIPP.

66. "Report: LANL contaminating nearby areas," *The Santa Fe New Mexican*, June 8, 1988, A-2, Fray Angélico Chávez History Library Vertical Files,
Santa Fe, NM: Palace of the Governors, Department of Cultural Affairs, Hazardous waste 1.

67. "WIPP Route Goes Through Heart of City."

68. Mike Taugher, "WIPP Bypass Roads Remain Incomplete," Center for

Southwest Research Vertical Files, Albuquerque: University Libraries, University of New Mexico, Energy—NM—Nuclear—WIPP.

69. Shonda Novak and Barbara Ferry, "WIPP Waste Inadvertently Crossed Sacred Tribal Land," *The New Mexican*, December 8, 1999, accessed March 25, 2015, http://www.newspapers.com/newspage/26732747/.

70. McCutcheon, *Nuclear Reactions*, 71, 79.

71. See "Summary of the Nuclear Waste Policy Act 42 U.S. C. §10101 et seq. (1982)," US Environmental Protection Agency, Washington, DC, November 12, 2014, accessed March 28, 2015, http://www2.epa.gov/laws-regulations /summary-nuclear-waste-policy-act.

72. Vincent B. Price, "A New Battle of Rival Beliefs Is Brewing over WIPP," February 16, 2001, D1, Center for Southwest Research Vertical Files, Albuquerque: University Libraries, University of New Mexico, Energy—NM—Nuclear—WIPP.

73. Richter, "New Mexico's Nuclear Enchantment," 12, 16, 38.

74. US Department of Energy, "Waste Isolation Pilot Plant Recovery Plan."

75. Jim Green, "WIPP Waste Accident a 'Horrific Comedy of Errors," *Nuclear Monitor*, no. 4430, November 20, 2014, accessed March 17, 2015, https:// www.wiseinternational.org/nuclear-monitor/794/wipp-waste-accident-hor rific-comedy-errors.

76. "IIFP Fluorine Extraction and Depleted Uranium Deconversion Plant Licensing," US NRC, Washington, DC, accessed December 16, 2014, http:// www.nrc.gov/materials/fuel-cycle-fac/inisfacility.html.

77. Elizabeth McNichol, Douglas Hall, David Cooper, and Vincent Palacios, "Pulling Apart: A State-by-State Analysis of Income Trends," Washington, DC: Economic Policy Institute and Center on Budget and Policy Priorities, November 15, 2012, accessed March 28, 2015, http://www.cbpp.org/cms/index.cfm ?fa=view&id=3860.

Conclusion

1. Ralph Vartabedian, "Goal of the Manhattan Project Historical Park Is to 'Remember and Learn from' Nation's Nuclear Achievements," *Los Angeles Times*, November 16, 2015.

2. Ben Hoyle, "'Disneyland for Nuclear Weapons' Divides America," *The Times*, November 19, 2015, https://www.thetimes.co.uk/article/ disneyland-for-nuclearweapons-divides-america-9qfpkstffml; Greg Mello and the Los Alamos Study Group, "The 'Manhattan Project National Historical Park': Moral Failure for America, Danger to This Country and the World," LASG.org, November 11, 2015, http://www.lasg.org/MPNHP/MPNHP_ brochure.pdf.

3. Hodge and Weinberger, *A Nuclear Family Vacation*, 2, 31, 35, 40, 50–51. Also see Vanderbilt, *Survival City*.

4. They include the Bradbury Science Museum, the Los Alamos Historical Society Museum, the National Museum of Nuclear Science and History, Grants' Uranium Mining Museum, the White Sands Missile Range Museum, and the Trinity site.

5. Hayashi, *From Trinity to Trinity*, xxiii, 51.

6. Szasz, *The Day the Sun Rose Twice*, 167.

7. See Palevsky, *Atomic Fragments*, 13.

8. Author's visit to the Bradbury Science Museum on November 23, 2012.

9. "Report from the Hilltop: Highlights of the Los Alamos Bradbury Science Museum," National Toxic Land/Labor Conservation Service, May 31, 2013, accessed March 1, 2014, http://www.nationaltlcservice.us/2013/05/los-alamos -bradbury-science-museum/. Also see Bartlit, "A Communication Analysis of Visitor Comments after Observing an Antinuclear and a Veterans' Exhibit in the Bradbury Science Museum of the Los Alamos National Laboratory."

10. Author's visit of the Bradbury Science Museum on November 23, 2012. These testimonies and pictures of each person, which were framed and hung in the museum, are extracted from Melnick, *They Changed the World*.

11. "Manhattan Project TV Series to Begin Production in New Mexico," *New York Post*, March 6, 2014, accessed April 22, 2016, http://nypost .com/2014/03/06/manhattan-project-series-to-begin-production-in-new -mexico/. Emphasis added.

12. Steve Terrell, "Manhattan Series Filmed in Sana Fe Focuses on Toll of Building the Bomb," *Santa Fe New Mexican*, June 21, 2014, accessed April 22, 2016, http://www.santafenewmexican.com/news/local_news/manhattan -series-filmed-in-santa-fe-focuses-on-toll-of/article_e2766a99-1851-5e5c-90 b5-9f30c24c90b5.html.

13. Sam Shaw, *Manhattan*, "Behold the Lord High Executioner," Season 2, Episode 7, November 24, 2015.

14. These include, among others, *The Beginning or the End*, by Norman Taurog (1947); *Fat Man and Little Boy*, by Robert Joffe (1989); *Day One*, by Joe Sargent (1989); the opera *Doctor Atomic* (2006), with a libretto by Peter Sellars and music by John Adams; and the play *In the Matter of J. Robert Oppenheimer* (2006), by Tom Morton-Smith.

15. The title was inspired by Jennifer Richter's dissertation on nuclear waste, and also by photographer Patrick Nagatani's work in the 1990s; see Patrick Nagatani, "New Mexico's Nuclear Enchantment," New Mexico PBS, September 30, 1991, accessed March 23, 2018, http://portal.knme.org/video /1476806820/.

Selected Bibliography

Archival material

Center for Southwest Research Vertical Files. Albuquerque: Center for Southwest Research. University Libraries. University of New Mexico.

Eidenbach, Peter L., Beth Morgan, and Mark Carter, eds. "Homes on the Range: Oral Recollections of Early Ranch Life on the U.S. Army White Sands Missile Range, New Mexico." US Department of Defense, Legacy Resource Management Program. Ranching Heritage Oral History Project. Las Cruces, NM: Human Systems Research, 1994.

"Farm and Ranch Folks Project." Oral History Program. Las Cruces, NM: New Mexico Farm and Ranch Heritage Museum, New Mexico Department of Cultural Affairs. Research and collections.

Ferenc Morton Szasz Papers, 1894–2005. Albuquerque: Center for Southwest Research. University Libraries. University of New Mexico. Collection MSS552BC.

Fray Angélico Chávez History Library Vertical Files. Santa Fe, NM: Palace of the Governors, Department of Cultural Affairs.

Governor John Burroughs Papers, 1959–1960. Santa Fe, NM: New Mexico State Records Center and Archive. Collection 1959–239.

Governor John Dempsey Papers, 1936–1958. Santa Fe, NM: New Mexico State Records Center and Archive. Collection 1959–046.

"Impact Los Alamos Project." Oral History Projects and Video Recordings Collection. Albuquerque: Center for Southwest Research. University Libraries. University of New Mexico. Collection MSS821BC, 1984–2006.

Malmgren, Peter. "Los Alamos Revisited, an Oral History." Santa Fe, NM: New Mexico State Records Center and Archives, 2000.

Ralph Carlisle Smith Papers on Los Alamos 1924–1957. Albuquerque: Center for Southwest Research. University Libraries. University of New Mexico. Collection MSS149BC.

"Voices of the Manhattan Project." Washington, DC: Atomic Heritage Foundation. Los Alamos, NM: Los Alamos Historical Society, 2012.

Reports

Adelamar, Alcantara, Kevin Kargacin, Marie Mora, Richard Santos, and Lawrence A. Waldman. "Poverty in New Mexico: Who Are the Poor?" Albuquerque: University of New Mexico, Bureau of Business and Economic Research, 1997.

Advisory Committee on Human Radiation Experiments. "The Human Radiation Experiments." New York: Oxford University Press, 1996.

Anschuetz, Kurt F., and Thomas Merlan. "More than a Scenic Mountain Landscape: Valles Caldera National Preserve Land Use History." General Technical Report RMRS-GTR-196. Fort Collins, CO: US Department of Agriculture, Forest Service, and Rocky Mountain Research Station, September 2007.

Bishaw, Alemayehu. "Poverty: 2010 and 2011 American Community Survey Briefs." US Department of Commerce, Economics, and Statistics Administration, September 2012.

Cohen, Timothy M. "The Economic Impact of Los Alamos National Laboratory and Sandia National Laboratories on the State of New Mexico Fiscal Year 1990." Las Cruces: New Mexico State University, College of Agriculture and Home Economics, Agricultural Experiment Station, Cooperative Extension Service, 1991.

Hoard, Dorothy. "Historic Transportation Routes on the Pajarito Plateau." LA-UR-06–3550. Los Alamos, NM: Los Alamos National Laboratory Ecology Group, Environmental Stewardship Division, May 2006.

Krens, D. L., and C. A. Sankey. "WIPP and the Local Communities." Albuquerque; Carlsbad, NM: US Department of Energy Albuquerque Operations Office; Waste Isolation Pilot Plant, 1986.

Krischner Associates, Management and Economic Consultants. "Adjustments to Reduced National Defense Expenditures in New Mexico: A Regional Economic Study for the United States Arms Control and Disarmament Agency." Contract No. ACDA/E-58 Summary Report. Albuquerque, December 1965.

McDonald, Brian. "The New Mexico Economy: History and Outlook." Albuquerque: University of New Mexico, Institute for Applied Research, Bureau of Business and Economic Research, 1988.

McNichol, Elizabeth, Douglas Hall, David Cooper, and Vincent Palacios. "Pulling Apart: A State-by-State Analysis of Income Trends." Washington, DC: Economic Policy Institute and Center on Budget and Policy Priorities, November 15, 2012.

New Mexico Energy Institute. "Attitudes of New Mexico Residents Toward the Nuclear Fuel Industry." NMEI Report No. 76–513A. Albuquerque: University of New Mexico, January 1977.

North Central New Mexico Economic Development District. "Regional Development Plan for the North Central New Mexico Economic Development District." Santa Fe, NM, June 1977.

Power, Thomas Michael. "An Economic Evaluation of a Renewed Uranium Mining Boom in New Mexico." Santa Fe: New Mexico Environmental Law Center, October 2008, accessed April 19, 2015, http://kenanaonline.com /files/0086/86224/NMUraniumEconomics[1].pdf

Purtymun, William D., and William R. Kennedy. "Geology and Hydrology of Mesita del Buey." LA-4660 UC-41 Health and Safety TIF-4500. Los Alamos, NM: Los Alamos Scientific Laboratory of the University of California, November 1970–May 1971.

Ross, Donald C. "Descriptive Petrography of Three Large Granitic Bodies in the Inyo Mountains, California." Geological Survey Professional Paper 601. Washington, DC: US Government Printing Office, 1969.

Rupasingha, Anil, and J. Michael Patrick. "Rural New Mexico Economic Conditions and Trends." CR-651. Las Cruces: New Mexico State University, College of Agricultural, Consumer and Environmental Sciences, Cooperative Extension Service, April 2012.

US Congress, Office of Technology Assessment. "After the Cold War: Living with Lower Defense Spending." OTA-ITE-524. Washington, DC: US Government Printing Office, February 1992.

US Department of Energy. "Waste Isolation Pilot Plant Recovery Plan." September 30, 2014, accessed December 16, 2014, http://www.wipp.energy.gov /Special/WIPP%20Recovery%20Plan.pdf.

US Department of the Interior. "Annual Report of the Secretary of the Interior for the Fiscal Year Ended June 30, 1938." Washington, DC: US Government Printing Office, 1938, available at https://archive.org/details/annual reportofse8231unit.

US National Research Council and Committee to Provide Interim Oversight of the DOE Nuclear Weapons Complex. "The Nuclear Weapons Complex: Management for Health, Safety, and the Environment." Washington, DC: National Academies Press, 1989.

Welsh, Michael. "Dunes and Dreams: A History of White Sands National Monument." Professional Paper No. 55. Santa Fe, NM: National Park Service, Division of History, Intermountain Cultural Resources Center, 1995.

Whan, Glenn A., chairman. "The State of New Mexico: Governor's Energy Task Force." Committee on Nuclear Energy. Santa Fe, NM: Executive Office of the Governor, March 18, 1975.

Widner, Thomas, et al. "Draft Final Report of the Los Alamos Historical Document Retrieval and Assessment (LAHDRA) Project." Atlanta: Centers for Disease Control and Prevention (CDC), National Center for Environmental Health Division of Environment Hazards and Health Effects, Radiation Studies Branch, June 2009.

Theses and Dissertations

Bartlit, Nancy R. "A Communication Analysis of Visitor Comments after Observing an Antinuclear and a Veterans' Exhibit in the Bradbury Science Museum of the Los Alamos National Laboratory." Master's thesis, University of New Mexico, 1998.

Bittman, Richard A. "Dependency and the Economy of New Mexico." Master's thesis, University of New Mexico, 1955.

Chambers, Marjorie Bell. "Technically Sweet Los Alamos: The Development of a Federally Sponsored Scientific Community." PhD diss., University of New Mexico, 1974.

Dietz, Chris. "The Impact of Los Alamos National Laboratory on Northern New Mexico." Master's thesis, New Mexico Highlands University, 1989.

Karafantis, Layne Rochelle. "Weapons Labs and City Growth: Livermore and Albuquerque, 1945–1975." Master's thesis, University of Nevada, 2012.

Meyer, Mary A. "Jobs, Work-Related Values, and Attitudes of Staff and Technicians at a National Laboratory." PhD diss., University of New Mexico, 1984.

Richter, Jennifer. "New Mexico's Nuclear Enchantment: Local Politics, National Imperatives, and Radioactive Waste Disposal." PhD diss., University of New Mexico, 2013.

Ross, Annie Grace. "One Mother Earth, One Doctor Water: A Story About Environmental Justice in the Age of Nuclearism. A Native American View." PhD diss., University of California, 2002.

Stergioulas, Nikolas Athanassios. "An Analysis of the Structural Causes of Poverty in New Mexico." Master's thesis, University of New Mexico, 1979.

Wood, Robert Turner. "The Transformation of Albuquerque 1945–1972." PhD diss., University of New Mexico, 1980.

Articles

Amundson, Michael A. "Home on the Range No More: The Boom and Bust of a Wyoming Uranium Mining Town, 1957–1988." *Western Historical Quarterly* 26 (Winter 1995): 483–505.

Churchill, Ward, and Winona LaDuke. "Native America: The Political Economy of Radioactive Colonialism." *Critical Sociology* 13, no. 3 (April 1986): 51–78.

Gray, Kristina Fisher. "Reclaiming Querencia: The Quest for Culturally Appropriate, Environmentally Sustainable Economic Development in Northern New Mexico." *Natural Resources Journal* 48, no. 2. (Spring 2008): 479–531.

Inkret, William C., John C. Taschner, and Charles B. Meinhold. "A Brief History of Radiation Protection Standards." *Los Alamos Science*, no. 23 (1995): 116–23.

Jones, Cynthia Gillian. "A Review of the History of U.S. Radiation Protection Regulations, Recommendations, and Standards." *Health Physics, The Radiation Safety Journal* 88, no. 2 (February 2005): 105–24.

Klein, Christine A. "Treaties of Conquest: Property Rights, Indian Treaties, and the Treaty of Guadalupe Hidalgo." *New Mexico Law Review* 26 (1996): 201–55.

Knowlton, Clark S. "Causes of Land Loss Among the Spanish Americans in Northern New Mexico." *Rocky Mountain Social Science Journal* 1 (May 1963): 202–11.

Massachusetts Institute of Technology. "Radiation, How Much Is Considered Safe for Humans?" *MIT News*. Cambridge, MA: MIT News Office. Accessed May 19, 2014. http://newsoffice.mit.edu/1994/safe-0105.

Masur, Louis P. "Bernard DeVoto and the Making of the Year of Decision: 1846." *Reviews in American History* 18, no. 3 (September 1990): 436–51.

McLemore, Virginia T., and New Mexico Bureau of Geology and Min. Res., Socorro, NM. "Uranium Resources in New Mexico." Society of Mining, Metallurgy & Exploration Annual Meeting, February 25–February 28, 2007.

Oppenheimer, J. Robert. "Crossing." *Hound & Horn: A Harvard Miscellany* 1, no. 4 (June 1928).

O'Sullivan, John. "The Great Nation of Futurity." *United States Democratic Review* 6, no. 23 (November 1839): 426–30.

Price, Vincent B. "Edith Warner in the Shadow of Los Alamos." Provincial Matters. *New Mexico Mercury*. August 19, 2013.

Rhode, Paul. "The Nash Thesis Revisited: An Economic Historian's View." *Pacific Historical Review* 63, no. 3 (August 1994): 363–92.

Richards, Linda M. "On Poisoned Ground." *Distillations* (Spring 2013).

Smith, Alice Kimball. "Scientists and Public Issues." *Bulletin of the Atomic Scientists* 38, no. 10 (December 1982): 38–45.

Southwest Crossroads Spotlight. "The Pajarito Plateau and Los Alamos." *Cultures and Histories of the American Southwest.* Santa Fe, NM: SAR Press and School for Advanced Research, 2006.

Torrez, Robert J. "New Mexico's Spanish and Mexican Land Grants." *New Mexico Genealogical Society.* Albuquerque: New Mexico Genealogical Society, 1997.

Turkevich, Anthony. "Assuring Public Safety in Continental Weapons Tests." AEC Thirteenth Semiannual Report. *Bulletin of the Atomic Scientists* 9, no. 3 (April 1953): 85–89.

Turner, Stephanie B. "The Case of the Zia: Looking beyond Trademark Law to Protect Sacred Symbols." *Yale Law School Legal Scholarship Repository.* Student Scholarship Papers. Paper 124. January 3, 2012.

Widner, Thomas E. "The World's First Atomic Blast and How It Interacted with the Jornada del Muerto and Chupadera Mesa." *Geology of the Chupadera Mesa Region.* 60th Field Conference. *New Mexico Geological Society Guidebook* (2009): 425–28.

Books

Alberts, Don E., and Allan E. Putnam. *A History of Kirtland Air Force Base, 1928–1982.* Kirtland Air Force Base, NM: 1606 ABW Office of History, 1982.

Alexander, Frederic C., Jr. *History of Sandia Corporation through Fiscal Year 1963.* Albuquerque: Sandia, 1963.

Alperovitz, Gar. *The Decision to Use the Atomic Bomb.* New York: Vintage, 1996.

Amundson, Michael A. *Yellowcake Towns: Uranium Mining Communities in the American West.* Mining in the American West Series. Boulder: University Press of Colorado, 2002.

Badash, Lawrence, Joseph O. Hirschfelder, and Herbert P. Broida, eds. *Reminiscences of Los Alamos, 1943–1945.* Dordrecht: D. Reidel, 1980.

Barrera, Mario. *Race and Class in the Southwest.* 2nd ed. Notre Dame, IN: University of Notre Dame Press, 1979.

Barrillot, Bruno. *Les irradiés de la République: Les victimes des essais nucléaires français prennent la parole.* Les Livres du GRIP. Paris: Editions Complexe, 2003.

Bartimus, Tad, and Scott McCartney. *Trinity's Children: Living Along America's Nuclear Highway*. New York: Harcourt Brace Jovanovich, 1992.

Berthier-Foglar, Susanne. *Les Indiens Pueblo du Nouveau-Mexique: De l'arrivée des conquistadors à la souveraineté des nations pueblo*. Pessac, France: Presses Universitaires de Bordeaux, 2010.

Bethe, Hans A. *The Road from Los Alamos: Collected Essays of Hans A. Bethe*. Masters of Modern Physics Series. Melville, NY: American Institute of Physics, 1991.

Bird, Kai, and Martin J. Sherwin. *American Prometheus: The Triumph and Tragedy of J. Robert Oppenheimer*. New York: A. A. Knopf, 2005.

Blackett, Patrick M. S. *Fear, War and the Bomb*. New York: Whittlesey House, 1949.

Boyer, Paul S. *By the Bomb's Early Light: American Thought and Culture at the Dawn of the Atomic Age*. 2nd ed. Chapel Hill: University of North Carolina Press, 1994.

Briggs, Charles L., and John R. Van Ness, eds. *Land, Water, and Culture: New Perspectives on Hispanic Land Grants*. Albuquerque: University of New Mexico Press, 1987.

Brode, Bernice. *Tales of Los Alamos: Life on the Mesa, 1943–1945*. Edited by Barbara G. Storms. Los Alamos, NM: Los Alamos Historical Society, 1997.

Brugge, Doug, Timothy Benally, and Esther Yazzie-Lewis, eds. *The Navajo People and Uranium Mining*. Albuquerque: University of New Mexico Press, 2006.

Calvin, Ross. *Sky Determines: An Interpretation of the Southwest*. Southwest Heritage Series. Santa Fe, NM: Sunstone, 2016.

Campbell, John M. *Magnificent Failure: A Portrait of the Western Homestead Era*. Stanford, CA: Stanford University Press, 2002.

Cantelon, Philip L., Richard G. Hewlett, and Robert C. Williams, eds. *The American Atom: A Documentary History of Nuclear Policies from the Discovery of Fission to the Present*. 2nd ed. Philadelphia: University of Pennsylvania Press, 1991.

Castro, Rafaela G. *Chicano Folklore: A Guide to the Folktales, Traditions, Rituals, and Religious Practices of Mexican-Americans*. New York: Oxford University Press, 2000.

Caufield, Catherine. *Multiple Exposures: Chronicles of the Radiation Age*. Chicago, IL: The University of Chicago Press, 1989.

Chávez, Thomas E. *An Illustrated History of New Mexico*. Albuquerque: University of New Mexico Press, 2002.

Church, Peggy P. *The House at Otowi Bridge: The Story of Edith Warner and Los Alamos*. Albuquerque: University of New Mexico Press, 1960.

Clarfield, Gerard H., and William M. Wiecek. *Nuclear America: Military and Civilian Nuclear Power in the United States, 1940–1980.* New York: Harper & Row, 1984.

Clark, Ira G. *Water in New Mexico: A History of Its Management and Use.* Albuquerque: University of New Mexico Press, 1987.

Conant, Jennet. *109 East Palace: Robert Oppenheimer and the Secret City of Los Alamos.* New York: Simon & Schuster, 2005.

Córdova, Gilberto Benito. *The 3 1/2 Cultures of Española.* Albuquerque: El Norte, 1990.

Córdova, Kathryn M. *Children of the Pajarito Plateau: Manuel Lujan Sr., Lorenzita Lujan and Their Descendants.* Albuquerque: Downtown Printing, 2007.

David, E. J. R., ed. *Internalized Oppression: The Psychology of Marginalized Groups.* New York: Springer, 2014.

DeBuys, William. *Enchantment and Exploitation: The Life and Hard Times of a New Mexico Mountain Range.* Albuquerque: University of New Mexico Press, 1985.

DeBuys, William, and Alex Harris. *River of Traps: A Village Life.* Albuquerque: University of New Mexico Press in association with the Center for Documentary Studies, Duke University, 1990.

DeGroot, Gerard J. *The Bomb: A Life.* Cambridge, MA: Harvard University Press, 2005.

De la Garza, Rodolfo O., Z. Anthony Kruszewski, and Tomás A. Arciniega. *Chicanos and Native Americans: The Territorial Minorities.* Englewood Cliffs, NJ: Prentice-Hall, 1973.

DeVoto, Bernard A. *The Course of Empire.* Boston: Houghton Mifflin, 1952.

Dickson, Paul. *Sputnik: The Shock of the Century.* New York: Walker, 2001.

Domenici, Pete V. *A Brighter Tomorrow: Fulfilling the Promise of Nuclear Energy.* Lanham, MD: Rowman & Littlefield, 2004.

Dunbar-Ortiz, Roxanne. *An Indigenous Peoples' History of the United States.* Boston: Beacon, 2014.

———. *Roots of Resistance: A History of Land Tenure in New Mexico.* Norman: University of Oklahoma Press, 2007.

Ebright, Malcolm. *Land Grants and Lawsuits in Northern New Mexico.* Santa Fe, NM: Center for Land Grant Studies, 2008.

Edgington, Ryan H. *Range Wars: The Environmental Contest for White Sands Missile Range.* Lincoln: University of Nebraska Press, 2014.

Eichstaedt, Peter H. *If You Poison Us: Uranium and Native Americans.* Santa Fe, NM: Red Crane, 1994.

Ermenc, Joseph J., ed. *Atomic Bomb Scientists: Memoirs, 1939–1945: Interviews* with Werner Karl Heisenberg, Paul Harteck, Lew Kowarski, Leslie R. Groves, Aristid Von Grosse, C. E. Larson.* Westport, CT: Meckler, 1989.

Etulain, Richard W., ed. *Contemporary New Mexico, 1940–1990.* Albuquerque: University of New Mexico Press, 1994.

Fanon, Frantz. *Les Damnés de la terre.* Paris: Éditions La Découverte/Poche, 2002.

Ferguson, Niall. *Civilization: The West and the Rest.* London: Allen Lane, 2011.

Fergusson, Erna. *New Mexico, A Pageant of Three Peoples.* Albuquerque: University of New Mexico Press, 1973.

Fisher, Phyllis. *Los Alamos Experience.* Tokyo: Japan Publications, 1985.

Fox, Sarah Alisabeth. *Downwind: A People's History of the Nuclear West.* Lincoln, NE: Bison Books, 2014.

Fraser, Steve, and Gary Gerstle, eds. *The Rise and Fall of the New Deal Order 1930–1980.* Princeton, NJ: Princeton University Press, 1989.

Furman, Necah S. *Sandia National Laboratories: The Postwar Decade.* Albuquerque: University of New Mexico Press, 1990.

Gardner, Richard M. *Grito! Reies Tijerina and the New Mexico Land Grant War of 1967.* Indianapolis: Bobbs-Merrill, 1970.

Gerber, Carl R., Richard Hamburger, and E. W. Seabrook Hull. *Plowshare.* Understanding the Atom series. Oak Ridge, TN: US Atomic Energy Commission, Division of Technical Information Extension, 1966.

Gibson, Toni M., and Jon Michnovics. *Los Alamos 1944–1947.* Images of America. Charleston, SC: Arcadia, 2005.

Goldschmidt, Bertrand. *Atomic Complex: A Worldwide Political History of Nuclear Energy.* La Grange Park, IL: American Nuclear Society, 1982.

Gómez, Laura. *Manifest Destinies: The Making of the Mexican American Race.* New York: New York University Press, 2007.

Gonzales, Raymond Bences. *A Boy on the Hill.* Edited by Judith Gursky. Los Alamos, NM: Los Alamos Historical Society, 2001.

Greene, Gayle. *The Woman Who Knew Too Much: Alice Stewart and the Secrets of Radiation.* Ann Arbor: University of Michigan Press, 1999.

Gregg, Josiah. *Commerce of the Prairies, or, the Journal of a Santa Fé Trader: During Eight Expeditions across the Great Western Prairies, and a Residence of Nearly Nine Years in Northern Mexico.* New York: J. & H. G. Langley, 1845.

Groves, Leslie R. *Now It Can Be Told: The Story of the Manhattan Project.* New York: Da Capo, 1983.

Gusterson, Hugh. *Nuclear Rites: A Weapons Laboratory at the End of the Cold War.* Berkeley: University of California Press, 1996.

Guthrie, Thomas H. *Recognizing Heritage: The Politics of Multiculturalism in New Mexico*. Lincoln: University of Nebraska Press, 2013.

Gutiérrez, David G., ed. *Between Two Worlds: Mexican Immigrants in the United States*. Wilmington, DE: Scholarly Resources, 1996.

Hacker, Barton C. *The Dragon's Tail: Radiation Safety in the Manhattan Project, 1942–1946*. Berkeley: University of California Press, 1987.

Haenn, Nora, and Richard Wilk, eds. *The Environment in Anthropology: A Reader in Ecology, Culture, and Sustainable Living*. New York: New York University Press, 2005.

Hales, Peter B. *Atomic Spaces: Living on the Manhattan Project*. Urbana: University of Illinois Press, 1997.

Hall, G. Emlen. *Four Leagues of Pecos, A Legal History of the Pecos Grant, 1800–1933*. Albuquerque: University of New Mexico Press, 1984.

Ham, Paul. *Hiroshima Nagasaki: The Real Story of the Atomic Bombings and Their Aftermath*. Perth: Black Swan, 2013.

Hamilton, Lisa M. *Deeply Rooted: Unconventional Farmers in the Age of Agribusiness*. Berkeley: Counterpoint, 2010.

Hanson, Rob E. *The Great Bisbee I. W. W. Deportation of July 12, 1917*. Signature Press, 1989.

Hasegawa, Tsuyoshi. *Racing the Enemy: Stalin, Truman, and the Surrender of Japan*. Cambridge, MA: Harvard University Press, 2005.

Hayashi, Kyôko. *From Trinity to Trinity*. Translated by Eiko Atake. New York: Midpoint Trade Books, 2010.

Hayward, John, ed. *John Donne: Complete Poetry and Selected Prose*. New York: Random House, 1949.

Herken, Gregg. *The Winning Weapon: The Atomic Bomb in the Cold War 1945–1950*. Princeton, NJ: Princeton University Press, 1981.

Hevly, Bruce W., and John M. Findlay, eds. *The Atomic West*. Seattle: University of Washington Press, 1998.

Hirt, Paul W., ed. *Terra Pacifica: People and Place in Northwest States and Western Canada*. Pullman: Washington State University Press, 1998.

Hodge, Nathan, and Sharon Weinberger. *A Nuclear Family Vacation: Travels in the World of Atomic Weaponry*. New York: Bloomsbury, 2008.

Hogan, Michael J., ed. *Hiroshima in History and Memory*. New York: Cambridge University Press, 1996.

Holdstock, Douglas, and Franck Barnaby, eds. *The British Nuclear Weapons Programme, 1952–2002*. London: Frank Cass, 2003.

Hunner, Jon. *Inventing Los Alamos: The Growth of an Atomic Community*. Norman: University of Oklahoma Press, 2004.

———. *J. Robert Oppenheimer, the Cold War, and the Atomic West*. Norman: University of Oklahoma Press, 2009.

Jacobson, Annie. *Operation Paperclip: The Secret Intelligence Program that Brought Nazi Scientists to America*. Boston: Little, Brown, 2014.

Jette, Eleanor. *Inside Box 1663*. Los Alamos, NM: Los Alamos Historical Society, 1977.

Jones, Vincent C. *Manhattan, the Army and the Atomic Bomb*. Washington, DC: Center of Military History, 1985.

Julyan, Robert. *The Place Names of New Mexico*. Albuquerque: University of New Mexico Press, 1998.

Jungk, Robert. *Brighter Than a Thousand Suns: A Personal History of the Atomic Scientists*. New York: Harcourt Brace Jovanovich, 1958.

Kay, Elizabeth. *Chimayó Valley Traditions*. Santa Fe, NM: Ancient City, 1987.

Kelly, Cynthia C., ed. *Remembering the Manhattan Project: Perspectives on the Making of the Atomic Bomb and Its Legacy*. Hackensack, NJ: World Scientific, 2004.

———. *The Manhattan Project: The Birth of the Atomic Bomb in the Words of Its Creators, Eyewitnesses, and Historians*. New York: Black Dog & Leventhal, 2007.

Kiernan, Denise. *The Girls of Atomic City: The Untold Story of the Women Who Helped Win World War II*. Reprint ed. New York: Touchstone, 2014.

Kuletz, Valerie L. *The Tainted Desert: Environmental Ruin in the American West*. New York: Routledge, 1998.

Kunetka, James W. *City of Fire: Los Alamos and the Atomic Age, 1943–1945*. Albuquerque: University of New Mexico Press, 1979.

Lamar, Howard R. *The Far Southwest, 1846–1912: A Territorial History*. Albuquerque: University of New Mexico Press, 2000.

Lamont, Lansing. *Day of Trinity*. New York: Athenaeum, 1965.

Lange, Charles H. *Cochiti: A New Mexico Pueblo, Past and Present*. Albuquerque: University of New Mexico Press, 1959.

Lazzell, Carleen, and Melissa Payne. *Historic Albuquerque: An Illustrated History*. San Antonio: Historical Publishing Network, 2007.

Leeds, Anthony, and Andrew P. Vayda, eds. *Man, Culture, and Animals: The Role of Animals in Human Ecological Adjustments*. Washington, DC: American Association for the Advancement of Science, 1965.

Lifton, Robert J., and Richard Falk. *Indefensible Weapons: The Political and Psychological Case Against Nuclearism*. New York: Basic Books, 1982.

Limerick, Patricia N. *The Legacy of Conquest: The Unbroken Past of the American West*. New York: W. W. Norton, 2006.

————. *Something in the Soil: Legacies and Reckonings in the New West.* New York: W. W. Norton, 2000.

Los Alamos Historical Society. *Behind Tall Fences: Stories and Experiences about Los Alamos at its Beginning.* Los Alamos, NM: Author, 1996.

Los Alamos Ranch School. *Los Alamos Ranch School.* Otowi, NM: The School, 1937.

Machen, Judith, Ellen McGehee, and Dorothy Hoard. *Homesteading on the Pajarito Plateau, 1887–1942.* Los Alamos, NM: Los Alamos National Laboratory, 2012.

MacLachlan, Colin M., and Jaime E. Rodríguez O. *The Forging of the Cosmic Race: A Reinterpretation of Colonial Mexico.* Berkeley: University of California Press, 1980.

Maclear, Kyo. *Beclouded Visions: Hiroshima-Nagasaki and the Art of Witness.* New York: State University of New York Press, 1998.

Magoffin, Susan S. *Down the Santa Fe Trail and into Mexico: The Diary of Susan Shelby Magoffin, 1846–1847.* Edited by Stella M. Drumm. American Tribal Religions, Book 3. Lincoln: University of Nebraska Press, Bison Books, 1982.

Martin, Craig. *Quads, Shoeboxes, and Sunken Living Rooms: A History of Los Alamos Housing.* Los Alamos, NM: Los Alamos Historical Society, 2000.

————. *Valle Grande: A History of the Baca Location No. 1.* Los Alamos, NM: All Seasons, 2003.

Masco, Joseph. *The Nuclear Borderlands: The Manhattan Project in Post–Cold War New Mexico.* Princeton, NJ: Princeton University Press, 2006.

Mason, Katrina R. *Children of Los Alamos: An Oral History of the Town Where the Atomic Age Began.* Twayne's Oral History Series, Book 19. New York: Twayne, 1995.

McCutcheon, Chuck. *Nuclear Reactions: The Politics of Opening a Radioactive Waste Disposal Site.* Albuquerque: University of New Mexico Press, 2002.

McKee, Robert E. *The Zia Company in Los Alamos: A History.* El Paso, TX: Carl Hertzog, 1950.

Melnick, Aj. *They Changed the World: People of the Manhattan Project.* Santa Fe, NM: Sunstone, 2006.

Melzer, Richard. *Coming of Age in the Great Depression: The Civilian Conservation Corps in New Mexico.* Las Cruces, NM: Yucca Tree, 2000.

Mills, Charles W. *The Power Elite.* New York: Oxford University Press, 2000.

Minear, Richard H., ed. *Hiroshima: Three Witnesses.* Princeton, NJ: Princeton University Press, 1990.

Mitchell, Pablo. *Coyote Nation: Sexuality, Race, and Conquest in Modernizing New Mexico, 1880–1920.* Chicago: University of Chicago Press, 2005.

Montaño, Chuck. *Los Alamos, Secret Colony, Hidden Truths: A Whistleblower's Diary*. Sante Fe, NM: Desert Tortoise, 2015.

Montoya, Maria E. *Translating Property: The Maxwell Land Grant and the Conflict over Land in the American West, 1840–1900*. Lawrence: University Press of Kansas, 2005.

Moss, Norman. *Men Who Play God: The Story of the Hydrogen Bomb and How the World Came to Live with It*. New York: Harper & Row, 1968.

Nash, Gerald D. *The American West Transformed: The Impact of the Second World War*. Lincoln: University of Nebraska Press, 1990.

———. *The Federal Landscape: An Economic History of the Twentieth-Century West*. Tucson: University of Arizona Press, 1999.

———. *World War II and the West: Reshaping the Economy*. Lincoln: University of Nebraska Press, 1990.

Nesbit, TaraShea. *The Wives of Los Alamos: A Novel*. New York: Bloomsbury, 2014.

Niklaus, Philip W. *How Safe is New Mexico's Atomic City? Radiation Control at Los Alamos Scientific Laboratory*. Albuquerque: Southwest Research & Information Center, 1980.

Ogura, Toyofumi. *Letters from the End of the World: A Firsthand Account of the Bombing of Hiroshima*. Translated by Kisaburo Murakami and Shigeru Fuji. Tokyo: Kodansha International, 1997.

Otero, Miguel Antonio. *My Life on the Frontier: 1864–1882*. Rev. ed. Santa Fe, NM: Sunstone, 2007.

Palevsky, Mary. *Atomic Fragments: A Daughter's Questions*. Berkeley: University of California Press, 2000.

Pasternak, Judy. *Yellow Dirt: An American Story of a Poisoned Land and a People Betrayed*. New York: Free Press, 2010.

Pomeroy, Earl S. *The American Far West in the Twentieth Century*. Edited by Richard W. Etulain. The Lamar Series in Western History. New Haven, CT: Yale University Press, 2008.

———. *The Territories and the United States, 1861–1890: Studies in Colonial Administration*. Philadelphia: University of Pennsylvania Press, 1947.

Powaski, Ronald E. *March to Armageddon: The United States and the Nuclear Arms Race, 1939 to the Present*. New York: Oxford University Press, 1987.

Price, Vincent B. *The Orphaned Land: New Mexico's Environment Since the Manhattan Project*. Albuquerque: University of New Mexico Press, 2011.

Rhodes, Richard. *The Making of the Atomic Bomb*. New York: Simon & Schuster, 1986.

Roberts, Calvin A., and Susan A. Roberts. *New Mexico*. Rev. ed. Albuquerque: University of New Mexico Press, 2006.

Rosenberg, Howard L. *Atomic Soldiers: American Victims of Nuclear Experiments.* Boston: Beacon, 1980.

Rothman, Hal. *Devil's Bargains: Tourism in the Twentieth-century American West.* Lawrence: University Press of Kansas, 1998.

———. *On Rims and Ridges: The Los Alamos Area Since 1880.* Lincoln: University of Nebraska Press, 1997.

Saad, Mohammed. *Development Through Technology Transfer: Creating New Organisational and Cultural Understanding.* Portland: Intellect Books, 2000.

Scarborough, Mike. *Trespassers on Our Own Land: Structured as an Oral History of the Juan P. Valdez Family and of the Land Grants of Northern New Mexico.* Indianapolis: Dog Ear, 2011.

Schumpeter, Joseph A. *The Theory of Economic Development: An Inquiry into Profits, Capital, Credit, Interest, and the Business Cycle.* Translated by Redvers Opie. Cambridge, MA: Harvard University Press, 1934.

Serber, Charlotte, and Jane Wilson, eds. *Standing by and Making Do: Women of Wartime Los Alamos.* Los Alamos, NM: Los Alamos Historical Society, 1988.

Serber, Robert. *Peace and War: Reminiscences of a Life on the Frontiers of Science.* New York: Columbia University Press, 1998.

Shroyer, Jo Ann. *Secret Mesa: Inside Los Alamos National Laboratory.* New York: John Wiley & Sons, 1998.

Simmons, Marc. *Albuquerque: A Narrative History.* Albuquerque: University of New Mexico Press, 1982.

———. *New Mexico: An Interpretive History.* Albuquerque: University of New Mexico Press, 1998.

———, ed. *On the Santa Fe Trail.* Lawrence: University Press of Kansas, 1986.

Slotkin, Richard. *Regeneration Through Violence: The Mythology of the American Frontier, 1600–1860.* Norman: University of Oklahoma Press, 2000.

Sparrow, James T. *Warfare State: World War II Americans and the Age of Big Government.* New York: Oxford University Press, 2011.

Spicer, Edward H. *Cycles of Conquest: The Impact of Spain, Mexico, and the United States on the Indians of the Southwest, 1533–1960.* Tucson: University of Arizona Press, 1992.

Steeper, Nancy C. *Gatekeeper to Los Alamos: Dorothy Scarritt McKibbin.* Los Alamos, NM: Los Alamos Historical Society, 2003.

Stephanson, Anders. *Manifest Destiny: American Expansionism and the Empire of Right.* Hill and Wang Critical Issue. New York: Hill & Wang, 1996.

Stoff, Michael B., Jonathan F. Fanton, and R. Hal Williams, eds. *The Manhattan*

Project: A Documentary Introduction to the Atomic Age. Philadelphia: Temple University Press, 1991.

Szasz, Ferenc M. *The Day the Sun Rose Twice: The Story of the Trinity Site Nuclear Explosion, July 16, 1945.* Albuquerque: University of New Mexico Press, 1984.

———. *Larger Than Life: New Mexico in the Twentieth Century.* Albuquerque: University of New Mexico Press, 2006.

Tehranian, Majid, ed. *Worlds Apart: Human Security and Global Governance.* New York: I. B. Tauris, 1999.

Tierney, Gail D., and Teralene S. Foxx. *Historical Botany of the Romero Cabin: A Family Homestead on the Pajarito Plateau.* Los Alamos, NM: Los Alamos National Laboratory, 1999.

Tijerina, Reies López. *They Called Me "King Tiger": My Struggle for the Land and Our Rights.* Houston: Arte Publico, 2000.

Turrentine, William Jackson. *Wagon Roads West: A Study of Federal Road Surveys and Construction in the Trans-Mississippi West, 1846–1869.* Berkeley: University of California Press, 1952.

Vanderbilt, Tom. *Survival City: Adventures Among the Ruins of Atomic America.* New York: Princeton Architectural Press, 2002.

Visgilio, Gerald R., and Diana M. Whitelaw, eds. *Our Backyard: A Quest for Environmental Justice.* Lanham, MD: Rowman & Littlefield, 2003.

Walker, J. Samuel. *Prompt and Utter Destruction: Truman and the Use of Atomic Bombs Against Japan.* Chapel Hill: University of North Carolina Press, 2004.

Warner, Edith. *In the Shadow of Los Alamos: Selected Writings of Edith Warner.* Edited by Patrick Burns. Albuquerque: University of New Mexico Press, 2001.

Weber, David J. *Spanish Frontier in North America.* New Haven, CT: Yale University Press, 2009.

Weigle, Marta, Frances Levine, and Louise Stiver, eds. *Telling New Mexico: A New History.* Santa Fe, NM: Museum of New Mexico Press, 2009.

Weinberg, Alvin M. *The First Nuclear Era: The Life and Times of a Technological Fixer.* New York: American Institute of Physics Press, 1994.

Welsome, Eileen. *The Plutonium Files: America's Secret Medical Experiments in the Cold War.* New York: Dial Press, 1999.

Westinghouse, George, ed. *Science and Civilization.* 3 vols. New York: McGraw-Hill, 1946.

White, Richard. *"It's Your Misfortune and None of My Own": A History of the American West.* Norman: University of Oklahoma Press, 1991.

White, Richard, and Patricia N. Limerick. *The Frontier in American Culture.*

Edited by James R. Grossman. Berkeley: University of California Press, 1994.

Wills, Garry. *Bomb Power: The Modern Presidency and the National Security State*. New York: Penguin, 2010.

Wirth, John D., and Linda K. Aldrich. *Los Alamos: The Ranch School Years, 1917–1943*. Albuquerque: University of New Mexico Press, 2003.

Zinn, Howard. *A People's History of the United States*. New York: HarperCollins, 2005.

Index

Page numbers in italic text indicate illustrations.